WEAPON

INFANTRY ANTIAIRCRAFT MISSILES

STEVEN J. ZALOGA
Series Editor Martin Pegler

Illustrated by Johnny Shumate & Alan Gilliland

OSPREY PUBLISHING
Bloomsbury Publishing Plc
Kemp House, Chawley Park, Cumnor Hill, Oxford OX2 9PH, UK
29 Earlsfort Terrace, Dublin 2, Ireland
1385 Broadway, 5th Floor, New York, NY 10018, USA
E-mail: info@ospreypublishing.com
www.ospreypublishing.com

OSPREY is a trademark of Osprey Publishing Ltd

First published in Great Britain in 2023

A catalog record for this book is available from the British
Library.

ISBN: PB 9781472853431; eBook 9781472853448;
ePDF 9781472853455; XML 9781472853424

23 24 25 26 27   10 9 8 7 6 5 4 3 2 1

Index by Rob Munro
Typeset by PDQ Digital Media Solutions, Bungay, UK
Printed and bound in India by Replika Press Private Ltd.

Osprey Publishing supports the Woodland Trust, the UK's
leading woodland conservation charity.

To find out more about our authors and books visit
www.ospreypublishing.com. Here you will find extracts, author
interviews, details of forthcoming events and the option to sign
up for our newsletter.

Author's note

During the Cold War (1947–91), the names and designations of
Soviet missiles were unknown to NATO. The US Weapon and
Space Systems Intelligence Committee (WSSIC; formerly the
Guided Missile and Astronautics Intelligence Committee
(GMAIC) until 1976) gives them an alphanumeric designation
such as SA-7 (Surface-to-Air-7). The multinational Air Force
Interoperability Council (AFIC; known as the Air
Standardization Coordination Committee prior to 2005) gives
them a name such as "Grail." These designations and names are
frequently linked together such as SA-7 "Grail" even though they
are separate designations and names given by separate
institutions. In this book, Soviet missiles are identified by their
Soviet/Russian name, though their archaic WSSIC/AFIC
designations are also identified.

The author would especially like to thank David Isby for his
generous help with photographs and information for this book;
and likewise to Wojciech Łuczak, editor of the defense magazine
Raport, for his help in obtaining photographs. Many of the
Afghanistan photographs come from the Afghan Media Resource
Center, which has deposited its extensive photograph collection
at the US Library of Congress for public use. Any photographs
not otherwise credited are from the author's collection.

Front cover, above: The 9K34 Strela-3 system. (Author)

Front cover, below: A Stinger team in training. (Author)

Title-page photograph: The 9K32M Strela-2M system was
designed for use from armored infantry vehicles on the move
and, in this example, from one of the rear hatches of a BMP-1
infantry fighting vehicle. This is a training example of the
Strela-2M and is actually missing a portion of the 9P53M
gripstock.

CONTENTS

INTRODUCTION

Ground-attack aircraft have posed a threat to land armies since World War I; but although machine guns were a deterrent against slow, propeller-driven aircraft, the advent of jet-powered aircraft during the 1940s made such weapons increasingly ineffective. Armies began to develop rocket antiaircraft weapons in World War II such as the German Fliegerfaust. Without guidance, however, these weapons had little chance of hitting a maneuvering attack aircraft.

The first attempts at developing a man-portable air defense system (MANPADS) employing a guided missile began in the mid-1950s. The first generation of MANPADS such as the US Army's Redeye and the Soviet Strela-2 (SA-7 "Grail") entered service in the late 1960s. The first generation of MANPADS were a technological offshoot of infrared (IR)-guided air-to-air missiles (AAMs) such as the AIM-9 Sidewinder. The principal challenge for these early MANPADS programs was to scale down the missile guidance seeker to a size small enough to fit in a man-portable missile. The first-generation guidance seekers consisted of a Cassegrain telescope that focused the IR energy emanating from a target aircraft on to an uncooled, photoconductive lead-sulfide detector. The resulting electrical signal was then processed by the onboard electronics to steer the missile toward the target using proportional guidance. (Proportional guidance means that the missile does not attempt to chase the target aircraft, but rather the flight-control system predicts where it expects the missile to intercept the target, and steers the missile to this point.)

Combat aircraft emit a wide range of IR radiation. The most radiant element on a jet fighter or gas-turbine helicopter is the hot jet exhaust and the hot metal of the exhaust pipe. The hot metal usually emits in the 2μm wavelength while the exhaust plume starts around 4μm and dissipates around 8μm. Other parts of the aircraft can also emit IR radiation, for example from aerodynamic heating of the aircraft skin, as well as sunlight

reflected off the aircraft surface. A United States Air Force (USAF) study offered a rough estimate of the relative IR signature of contemporary aircraft. If the IR signature of a helicopter such as the AH-64 Apache is given as 1, a turboprop transport such as the C-130 Hercules is 10, a tactical jet fighter such as an F-16 Fighting Falcon is 35, and a large jet transport such as a C-17 Globemaster III is 100.

There were two principal problems with the early MANPADS that used uncooled lead-sulfide detectors. Because these seekers were sensitive only in the 1.5–2.2μm band, they could only detect the hot metal of the jet exhaust pipe. Other IR radiation emanating from the aircraft, such as the dispersed exhaust gases, was invisible to these seekers. As a result, these early MANPADS could only fire at aircraft when the jet exhaust pipes were visible. In the case of jet fighters, this meant when they were passing away from the MANPADS with their tails exposed. The MANPADS could, in limited cases, target crossing targets such as helicopters with side-mounted exhausts, or jets with particularly prominent exhausts.

The second problem was that these seekers were extremely vulnerable to being overwhelmed by solar radiation. The MANPADS missiles were useless if the target aircraft had the sun behind them because solar radiation was far more radiant than the jet exhaust. The uncooled detectors could also be misdirected by sun-glint, especially sun-glint reflected off clouds.

The first Soviet MANPADS, the Strela-2, used uncooled lead-sulfide detectors in spite of their acknowledged limitations. Soviet specifications estimated that this system had a probability-of-hit of only about

Most MANPADS use a small booster to eject the missile from the launch tube to minimize exhaust blast against the gunner. The main missile rocket motor ignites away from the launcher, as shown here during a FIM-92 Stinger test launch at Eglin Air Force Base, Florida, in 2014. (US Air Force, 96th Test Wing Public Affairs, Samuel King Jr.)

25 percent. In practice, it was even lower than this. While this performance may seem inadequate to justify the expense of such weapons, two tactical considerations justified their deployment.

First, the performance of the early MANPADS, although far from ideal, was significantly better than previous types of weapons such as heavy machine guns (HMGs) or autocannons. It was estimated that it would take 8,500 rounds fired from the Soviet 57mm S-60 towed antiaircraft gun or ZSU-57-2 self-propelled antiaircraft gun (SPAAG) to have a high probability of shooting down a jet fighter. Even the radar-directed 23mm ZSU-23-4 SPAAG was estimated to have a probability-of-kill against a relatively slow-flying helicopter of only 18 percent at a 2km range.

Second, a missile did not actually have to shoot down an enemy aircraft in order to fulfill its air defense mission. Although a shoot-down was the ideal outcome, hindering an enemy aircraft from making an accurate attack against ground targets was still an acceptable outcome. In air defense doctrine, this is known as "virtual attrition": although the enemy aircraft is not actually "attrited" by being shot down, it can be "virtually attrited" if it is prevented from fulfilling its mission. MANPADS cause virtual attrition by forcing enemy attack aircraft to break off an attack, or by forcing them to attack from higher altitudes from which they have a lower probability of hitting their target.

The first combat use of a MANPADS was the employment of the Soviet Strela-2 by Egyptian forces in the War of Attrition against Israeli aircraft in August 1969 (see page 47).

Although early prototypes of the US Army's Redeye MANPADS used an uncooled detector, innovations in seeker design led to the advent of the first cooled detectors. The initial technology was the thermoelectrically cooled detector, often based on the same lead-sulfide array as the early uncooled types. Although these cooled detectors had some advantages over uncooled detectors such as greater sensitivity, they tended to be limited to the 2.6–2.7μm band. This type of detector had advantages over other first-generation seekers, but it still used many of the same technologies, relying on reticle chopping and AM modulation for its sensor processing. This type of detector was used as an expedient in some stopgap MANPADS designs such as the short-lived XFIM-92A Redeye in 1966. The most significant MANPADS using a thermoelectrically cooled detector was the Soviet Strela-2M, starting in 1970. The Strela-2M and its various variants and foreign copies, manufactured in several countries and deployed by over 60 armies, was the most widely produced MANPADS of all time, with total production approaching a half-million missiles.

The problems associated with lead-sulfide detectors using reticle-chopping processing became far worse in the early 1970s, however, when air forces began to adopt IR countermeasures (IRCM). There were two early forms of IRCM, the simplest of which was an exhaust diffuser. These were especially common on helicopters, being used to duct the engine's hot exhaust gas up into the propeller wash, thus cooling and diffusing the exhaust gas. The first use of these devices was the US Army's

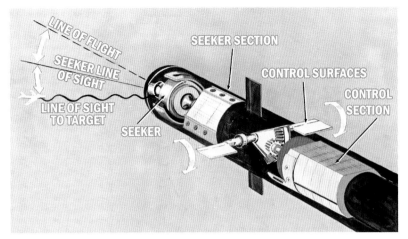

improvised "Toilet Bowl" diffuser for its helicopters in Vietnam in 1971 in response to the introduction of the Strela-2 there; the Soviet Air Force introduced similar diffusers in Afghanistan during 1984–85.

The second method of IRCM was IR flares, originally burning MTV (magnesium/Teflon/viton). Aircraft could be fitted with a dispenser that ejected dozens of small flares, emitting IR radiation in the same wavelengths as the detector sensitivity of the MANPADS. Because the flares were hotter than the aircraft, the flares would lure the MANPADS missile away from the aircraft.

The next step in the "Wizard War" was a new form of IRCM: modulated IR jammers. Nicknamed "hot bricks," these emitted modulated IR pulses that were synchronized with the rotation rate of MANPADS' seeker reticles. This prevented the seeker from locking on to the aircraft, or broke the lock-on if initiated after the missile had been fired. The first of these was the United States' AN/ALQ-144, which made its debut in 1981.

The shortcomings of uncooled seekers were overcome by using cryogenic cooling of the detector. For MANPADS, this usually meant the use of a pressurized gas such as nitrogen. The nitrogen was contained in a battery/coolant unit (BCU) that plugged into the gripstock launcher before an engagement. Prior to launch, the gunner would activate the BCU, which provided electrical power to the missile, uncage the seeker gimbal, which permitted it to swivel, and cool the detector with the pressurized gas. In 1967, the FIM-92C Redeye was the first mass-produced MANPADS to make use of cooled-detector technology, while the first Soviet MANPADS with a cooled seeker was the Strela-3 in 1974.

The next evolution in MANPADS introduced filters and guidance innovations to reduce the seeker's vulnerability to flares. These innovations were typically based on conical scanning and FM modulation for their signal processing. These technologies also reduced the seeker's vulnerability to being distracted by natural IR distractions such as sunlight. This generation of seekers also saw a shift from lead-sulfide detectors to indium antimonide. They were sensitive in the 3–5μm band, which enabled them

The classic countermeasure against the early MANPADS was the IR flare. These were more radiant than the aircraft's exhaust and so would lure away the early first-generation MANPADS guidance seekers. This C-5M Super Galaxy can be seen dispensing flares over the Air Force Operational Test and Evaluation Center at Eglin Air Force Base in May 2021. (US DoD, Samuel King, USAF 96th Test Wing Public Affairs, Eglin AFB)

to detect a more useful range of IR emissions from target aircraft including the exhaust plume, and engine heat on the fuselage sides.

This new generation of MANPADS also benefited from advances in microprocessor technology, epitomized by the Intel 4004 chip, which appeared in 1971 and permitted much more sophisticated processing techniques. Furthermore, the advance in chip technology was exceptionally rapid: the Intel 4004 chip of 1971 was capable of 92k operations per second; the Intel 8086 chip of 1978 could conduct 710k operations per second. As a result, this new generation of MANPADS was no longer limited to "tail-chasers," but could attack the enemy aircraft from multiple angles depending on the circumstances. Typical examples of this generation of MANPADS were the US FIM-92A Stinger and Soviet Igla.

The limitations of IR-guided MANPADS prompted Britain to consider other guidance alternatives for its Blowpipe missile. The most obvious was to use manual command-to-line of sight (MCLOS), a guidance method used in some early air-to-air and surface-to-air missiles as well as some antitank missiles. The usual command link for MCLOS guidance was a trailing wire, but this was impractical for a MANPADS. A more suitable alternative was a radio-command link. On firing the missile, the gunner kept both the target and missile in sight and steered the missile to the target using a small thumb controller. At least on paper, such a guidance approach made it possible to engage an enemy aircraft from any angle, including from the front.

The MCLOS guidance method proved to be ill-conceived. Guidance of the missile became progressively more difficult at longer ranges, and the gunner had to maintain guidance during the whole engagement. This paralleled similar problems with other first-generation MCLOS systems such as antitank missiles. In contrast, the IR-guided MANPADS were "fire-and-forget." Britain's military did not abandon the concept, however,

One of the earliest countermeasures applied to the MANPADS threat was exhaust diffusers. This is a Soviet EVU exhaust diffuser on an Mi-8AMTSh transport helicopter. The EVU drew in cool outside air on the front of the assembly and mixed it with the hot exhaust gases from the engine exhaust port behind it. This was expelled upward through the rear upper grilles into the rotor wash to diffuse and cool the exhaust heat signature. (Author)

but tried to redeem it using semi-automatic command-to-line of sight (SACLOS) on the follow-on Javelin missile. This reduced the gunner's burden by introducing an optical tracker in the launcher. The gunner simply had to keep the crosshairs on the target, and the launcher's tracker automatically transmitted the guidance corrections to the missile.

Another limitation of early MANPADS was their relatively low lethality due to the small size of the warhead. The French Mistral offered a more substantial missile, but at the cost of portability. The Mistral was too heavy to be launched from the shoulder, instead being launched from a pedestal. The French argument was that this reduction in portability was tactically irrelevant because most MANPADS squads are deployed on a vehicle anyway due to the weight of the missile, missile reloads, and communication equipment.

The third generation of IR-guided MANPADS such as the later versions of the US Stinger and Soviet Igla continued to evolve to defeat the various new forms of IRCM. The new semi-imaging or pseudo-imaging seekers of the 1980s used a detector with a sophisticated optical scanning process such as rosette scanning. These seekers also offered multispectral detection, adding ultraviolet (UV) detection beyond the usual IR wavelengths, to defeat simple IRCM tactics such as the use of flares.

There were other innovations beyond the missile itself. One major tactical problem associated with the use of MANPADS was that the operators had a difficult time distinguishing between friendly and hostile aircraft. The first method was to issue MANPADS units with simple display devices that were linked by radio to the army's air defense network. Radars and other sensors in the network passed their data to a central command post that in turn transmitted this data to MANPADS display devices. The first of these was the TADDS used with the Redeye MANPADS.

Another approach developed by the Soviet Union was the PRP (*passivniy radiopelengator*; passive radio directional alert device), a small direction-finding antenna worn on the gunner's helmet that detected the radar and other radio-frequency emissions from approaching enemy

Another approach to IRCM was the "hot brick," which contained a xenon lamp with a rotating deflector that modulated the IR signal to confuse the MANPADS guidance seeker. The SOEP-V1A Lipa was the initial Soviet "hot brick" IR jammer, introduced in 1985 in Afghanistan on platforms such as the Mi-24 "Hind" attack helicopter shown here. (Author)

aircraft. The PRP was sometimes used on the Strela-2M, but the technique was not especially successful and was eventually abandoned.

A more robust alternative was IFF (Identification Friend or Foe; svoy-chuzoy in Russian). This system relied on a transponder fitted to friendly aircraft that could be interrogated by the IFF located on the MANPADS. The aircraft transponder broadcast a coded signal that the IFF system on the MANPADS identified as friendly. These IFF systems were first used on the US Stinger and Soviet Igla. Contemporary MANPADS such as the Russian Verba are produced as a modular system that includes the usual missile and gripstock, as well as a network of IFF interrogators, display tablets, early-warning radars, and mobile air defense command posts.

The ability of advanced MANPADS such as the Stinger and Igla to circumvent older IRCM tactics led to the development of more sophisticated systems such as missile warning receivers. These devices detect the launch and approach of MANPADS missiles. They can be programmed to deploy flares automatically, and they warn the aircraft or helicopter crew so that evasive maneuvers can be initiated in time. Another innovation in countermeasures is the Directed Infrared Countermeasures (DIRCM), a turreted system that can be aimed at an approaching missile. A modulated laser beam is directed at the approaching missile's seeker, preventing the missile from maintaining a lock on the aircraft. First applied to large transport aircraft, these systems are being scaled down to permit mounting on strike aircraft and helicopters, but are only beginning to be deployed widely at the time of writing.

Owing to the sheer number of MANPADS developed over the past half-century, this book centers on the most important types that entered series production. The focus is on man-portable configurations; many of these missiles have specialized launchers mounted on vehicles, warships, and aircraft but these applications are not detailed here. The coverage of combat use of MANPADS focuses on conflicts that were especially important in their development or that made very extensive use of them. There have been many other conflicts in which MANPADS have seen small-scale use, but they are not explored in detail here due to lack of space.

DEVELOPMENT
A new antiaircraft defense

PRIMITIVE BEGINNINGS

A number of countries developed unguided antiaircraft rockets during World War II, but these mostly required large and heavy launchers. Germany was the only country to field a man-portable weapon, initially called the Luftfaust ("Air fist") but subsequently renamed Fliegerfaust ("Aircraft fist").

Hugo Schneider AG (HASAG) in Leipzig began development of the Luftfaust in July 1944 under the direction of the Heer's (German Army) Waffenprüfamt (Weapons Development Department). The requirement was for a disposable weapon like the Panzerfaust ("Tank fist") antitank rocket launcher. The munition was a standard 20mm projectile weighing 90g, including 19g of high explosive. Instead of the usual brass propellant case as used in a conventional gun, a small rocket motor with stabilizing fins was attached at the rear of the projectile. The requirement was for a munition speed of 300m/sec, a range of 500m, and a dispersion of no more than 10 percent at 500m. The Luftfaust consisted of four launch tubes stacked vertically, with a trigger and aiming sight derived from the Panzerfaust.

Initial tests of the weapon were disappointing because it was significantly overweight, had too short a range, and suffered from excessive dispersion. This led to a complete redesign of the weapon as the Luftfaust B in late 1944. Instead of being disposable, the Luftfaust B was reusable. The new configuration consisted of eight launch tubes mounted around a ninth in the center. When the weapon was fired, four projectiles were launched followed by a salvo of the remaining five after a 0.1-second delay. The weapon weighed 6.5kg (14.3lb) when loaded.

The Fliegerfaust was a primitive forerunner of the MANPADS, firing a spray of nine 20mm rocket-powered projectiles. It was deployed in the final days of World War II but there are no details of any combat use. This particular example is preserved at the Central Russian Armed Forces Museum in Moscow. (Author)

11

The 20mm munition for the Luftfaust B was also redesigned. The new rocket motor was longer, heavier, and consisted of a double charge, the first of which ejected the projectile from the launch tube, immediately followed by the ignition of the main rocket charge. Instead of using fins for stabilization as on the initial version, the rocket exhaust was vented through four canted venturi that spun the projectile at 26,000rpm for stabilization. The ammunition was issued in a nine-round clip to permit rapid loading.

In November 1944, the Wehrmacht decided to issue the Luftfaust B on a scale of 600 per division so that each rifle or machine-gun section would have such a weapon for air defense. Production began in January 1945 of 100 "0-Serie" preproduction launchers, and 1,000 clips of ammunition. About 80 of the launchers were earmarked for a trials unit in Saarbrücken, with troop trials beginning on March 15, 1945. In February 1945, the Luftfaust B was renamed Fliegerfaust. No details are available about any combat use of the Fliegerfaust, but its technical performance is known to have been disappointing because its effective range was often well under 500m. Dispersion remained far worse than desired: about 40m at 200m range.

Besides the Saarbrücken trials unit, Nazi politician Gauleiter (Regional Leader) Martin Mutschmann obtained a few Fliegerfaust weapons to defend his hunting lodge at Grillenburg in the Thrandt Forest, near the HASAG plant. Two examples were discovered at Grillenburg in 2004. There are also photos of several of these weapons in the rubble near Hotel Aldon opposite the Brandenburg Gate in Berlin in April 1945. Serial production was ordered in March 1945 for 10,000 launchers and 4 million projectiles, but it is unclear if any more were produced. A five-barrel, 30mm version was under development when World War II ended. Both the US Army and the Red Army captured small numbers of Fliegerfaust weapons, but the short range and wild dispersion of the rocket-powered projectiles did not encourage any further development after the war except for the short-lived Soviet Kolos (Spike) mentioned below.

THE PIONEER: REDEYE

In 1946, the US War Department Equipment Board determined that the US Army's primary short-range antiaircraft weapon, the .50-caliber HMG, would become ineffective against future threats due to its short range. The US Army sought a weapon that could be used on a stationary mounting, on vehicle mounts, and on a self-propelled antiaircraft vehicle. This program was called Stinger and was based around a new .60-caliber HMG. The program ended in 1951, however, when it was realized that the weapon could not meet the objective of a 14,000yd slant range. It was succeeded for a short time by a version based on a 37mm revolver cannon, but this proved too heavy and complex for the requirement and so was short-lived. It was followed by Project Octopus, a modular gun mounting that could be used with the .50- and .60-caliber HMGs and 20mm cannon. An alternative was the Porcupine, consisting of 64 launch tubes for 2.75in unguided rockets. Both these programs ended in 1956/57 because they were becoming too heavy, complex, and expensive for the short-range requirement.

In 1955, the Convair Division of General Dynamics Corporation began studying the feasibility of a small, IR-guided missile. IR missile guidance was already a proven technology that had been used on the AIM-4 Falcon and AIM-9 Sidewinder air-to-air missiles. The challenge was to develop a ground-based derivative that was compact and light enough to be carried by a foot soldier. In 1956, Convair built a full-scale model of the ground-based missile, nicknamed Redeye, due to its use of IR guidance. The main challenge was the seeker because the missile could use solid-rocket propulsion similar to that of the existing 2.75in unguided rocket. The initial version of the Redeye used a scaled-down version of the AIM-9 Sidewinder missile seeker.

Early prototypes of the Redeye were very small, as can be seen from this display of a mock-up in 1959. The design inevitably became larger and heavier as development progressed. (US Army)

The specifications called for a missile weighing 14.5lb and a gripstock/launch tube weighing 3.2lb for an overall weapon weight of 18.2lb. The launch system consisted of a disposable transport container/launch tube containing the missile, and a reusable gripstock that was attached to the launch tube and contained both the system power supply/coolant and the trigger system for firing the missile. The missile included a 1.2lb high-explosive warhead triggered by an impact fuze. As a result, the missile had to impact the enemy aircraft directly in order to detonate. The estimated unit cost was about $700, compared to about $3,000 for the much larger AIM-9 Sidewinder.

In late 1956, the US Army's main missile development center at Redstone Arsenal in Huntsville, Alabama, received a briefing on the proposed weapon. There was considerable enthusiasm for the concept. In early 1957, Redstone Arsenal solicited competitive industry bids for a "man-portable, all-arms weapon system" with the requirements very similar to Convair's Redeye concept. Two other firms bid for the development contact: Sperry Gyroscope with its Lancer and North American Aviation with its SLAM (Shoulder-Launched Antiaircraft Missile). The US Army rejected both the Lancer and SLAM as being too heavy. US Army engineers felt that Convair's Redeye offered the most promise, but that many basic engineering aspects needed substantial research. United States Marine Corps officers were more enthusiastic, however, and offered initial funding to accelerate the program. The US Army awarded Convair a feasibility study contract in January 1958 as the first step in the Redeye's development.

In the event, the US Army evaluation proved to be prescient. Development proved to be considerably more difficult than predicted, stretching to seven years, with the research costs tripling and the weapon's weight growing from 18.2lb to 29.3lb.

Initial feasibility tests were conducted at Naval Ordnance Test Station China Lake, California, which had developed the original AIM-9 Sidewinder missile. Tests of unguided Redeye missiles began in June 1958 and guided test launches in March 1959. The test launches proved the basic concept and the engineering development phase of the program began in October 1960. Convair contracted Philco Corporation to develop the seeker and Atlantic Research Corporation for the two-stage rocket motor.

Early tests disclosed numerous technical problems. One of the most protracted was controlling the roll rate of the missile in flight, involving not only the propulsion system, but also the pop-out tail fins. Performance of the initial uncooled lead-sulfide IR detector proved disappointing, leading to the start of two alternative approaches: an electrically cooled lead-sulfide detector and a Hughes cryogenically cooled lead-sulfide detector. The uncooled detectors were sensitive in the 2–2.7μm band, which could sense the high temperature of a jet aircraft's tailpipe. As a result, missiles using these first-generation uncooled detectors were essentially tail-chasers because they could not lock on to other aspects of the aircraft. The second-generation cooled detectors could lock on to the cooler portions of the jet exhaust plume as well as other IR-emitting portions of the aircraft. As a result, they had some ability to attack aircraft targets from wider angles than simply the tail. The US Army eventually decided to eschew uncooled detectors in favor of the next-generation cooled detectors.

Another change introduced during the Redeye's development was the launcher. The interface between the detachable gripstock and launcher tube proved to be a source of failures, leading to the decision to manufacture the system with both elements combined. Once fired, the unified gripstock/launch-tube could be recycled and reloaded at depots up to eight times.

The Redeye used a small booster on the tail of the missile to eject it out of the tube. The booster burned for only 60 milliseconds, and so did not emit any exhaust gases when it left the launch tube. The booster accelerated the missile to about 80ft/sec and propelled it about 25ft from the launch tube. This safeguarded the Redeye gunner from being injured by rocket exhaust. The main sustainer motor ignited at this point, and provided about 250lb of thrust for 5.6 seconds. The roll rate of the missile was controlled by the pop-out tail fins and the canted booster nozzles.

The continuing problems during test launches finally began to lessen in 1963. During Fiscal Year 1963 there were a total of 75 launches of which 23 were fully guided. The first successful hit on a QF-9F Panther jet aircraft drone flying at 275 knots (315mph) was made on December 14,

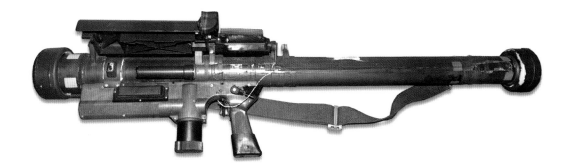

1962, followed by three more hits over the next few months. In October 1963, 13 missiles incorporating the improved features were fired at drones, with 11 direct impacts and the remaining two within 1ft of the thermal source. These successes convinced the US Army to begin the industrialization phase of the program.

The initial version of the production missile was the XM41 Block I weapon system that included the XFIM-43A missile with the Mod 60 thermoelectrically cooled detector, and the XM147 launcher. Delivery of over 300 XM41 Block I weapon systems began in September 1965 and ended in May 1966. These were expended in engineering and service tests. This was followed by the XM41E1 Block II weapon system that included the improved XFIM-43B missile with the Mod 60A cryogenically cooled detector. Delivery of the XM41E1 Block II weapon systems began in April 1966 alongside the final XM41 Block I missiles. In February 1967, the first XM41E1 Block II weapon systems were issued to US Army units for initial training. A total of 1,743 XFIM-43B missiles were manufactured and they were used primarily for engineering tests and troop training.

The M41 Block III weapon system was a major redesign of the Redeye and the first version put into actual US Army service. It used the new M171 launcher that could be easily distinguished from the previous XM147 by virtue of its new folding optical sight. The FIM-43C missile incorporated the Mod 60A cryogenically cooled detector and had all-new subcomponents including a new M222 warhead and improved M115 rocket motor. Production of Redeye weapon systems switched from XM41E1 Block II to M41 Block III in May 1967. Deliveries of M41 Block III weapon systems to US Army and US Marine Corps units began in March 1968, but it was not cleared for troop use in extreme climates until October 1968. The M41 Block III weapon system was formally standardized on December 18, 1968, roughly six years behind the original schedule.

The FIM-43C's seeker was sensitive enough to engage propeller-driven light aircraft such as the L-19/O-1 Bird Dog observation aircraft, but it was more effective when used against jet aircraft due to the greater radiant energy of jet engines. The overall probability-of-hit for Redeye was 30 percent against high-performance, maneuvering jet aircraft and 50 percent against slower-moving targets such as helicopters. Performance varied also due to range and other factors. Demonstrated performance

Unlike later IR-guided MANPADS, the FIM-43C Redeye was delivered with the gripstock and launch tube permanently connected. The sighting mechanism at the front of the gripstock was folded down when stored, but is shown here elevated for firing. (Author)

During the Cold War (1947–91), the US Army's Redeye teams typically deployed on a utility vehicle such as this M151 ¼-ton 4×4 truck. This enabled the team to carry additional missile rounds as well as support equipment such as a radio. (US Army)

was a 51 percent hit probability against a QF-9F Panther drone flying at 100m (328ft) altitude at 430 knots (495mph).

Redeye production for the US Army continued through 1969; foreign military sales started in the early 1970s, extending production to 1973. Other countries received Redeye missiles in later years, but these came from existing US Army stocks. In total, 6,639 Redeye missiles were exported.

FIM-43 REDEYE PROCUREMENT	
US Army	20,755
US Marine Corps	7,637
US Air Force	9
Australia	216
Denmark	540
West Germany	1,018
Sweden	1,093
Total	31,268

RED ARROW: THE STRELA-2, 2M, and 3

The second MANPADS to enter development was the Soviet Strela-2. The Russian term for this type of missile is PZRK (*Perenosnoy zenitniy raketniy kompleks*; portable antiaircraft missile system). The Soviet program was initiated in the late 1950s after the Soviet Army learned of the US development of the Redeye based on press and television accounts. The program was initiated in 1958 by GAU (redesignated GRAU on November 19, 1960). There was some suspicion that Soviet espionage played a role in this program due to the close similarity of many features of the Strela-2 with those of the Redeye.

The Soviet premier, Nikita Khrushchev, was an enthusiastic proponent of missile technology, and he directed that traditional arms-development bureaus begin to shift their attention to missiles. In the case of the Soviet MANPADS program, this mainly involved the Soviet small-arms industry. This industry had already been tasked with developing guided antitank missiles, and so the MANPADS program was an extension of this effort. Several different design bureaus were instructed to begin examining this requirement. GRAU selected Boris Ivanovich Shavyrin, chief designer at the Kolomna special design bureau (SKB-GA), to head this effort. Shavyrin had designed mortars during World War II and his design bureau was responsible for one of the first Soviet guided antitank missiles, the 3M6 *Shmel* (AT-1 "Snapper"), adopted in 1960. When Shavyrin died in 1965, his place was taken over by Sergey Pavlovich Nepobedimy, who would head subsequent Soviet MANPADS programs.

The Soviet small-arms industry had little experience in optical missile guidance, and so turned to the two Leningrad institutes for the development of the seeker. The seeker program was headed by O.A. Artamonov of the OKB-357 of the Leningrad Optical Industry Organization (LOMO) with a parallel program undertaken by G.A. Goryachin's design team at the State Optical Institute (GOI). This development effort was based on the development of the first Soviet IR-guided air-to-air missile, the K-5 (AA-1 "Alkali"), as well as the Soviet effort to reverse-engineer a captured AIM-9 Sidewinder air-to-missile obtained from China, which resulted in the K-13 (AA-2 "Atoll"). The Soviet approach was the same as the initial Redeye, using an uncooled lead-sulfide detector.

The overall technical/tactical requirement for the new missile was completed in 1960 for two IR-guided antiaircraft missiles: the larger,

The 9K32 Strela-2 system featured the 9P53 gripstock and 9P54 launch tube. Next to the launcher is the 9M32 missile that was delivered in a sealed 9P54 launch tube. This example, probably from Vietnam, is currently on display at the Udvar-Hazy Center of the National Air and Space Museum in Chantilly, Virginia. (Author)

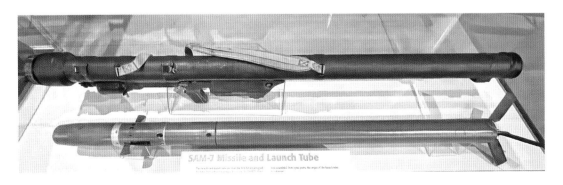

REVIVING THE FLIEGERFAUST: THE KOLOS

The problems with the Strela-2 led to a curious attempt to revive the German Fliegerfaust of 1945. In June 1966, TsNIITochmash (Central Scientific Research Institute of the Precision Machinery Industry) in Klimovsk received an order to develop a simple and inexpensive antiaircraft weapon patterned after the German weapon, codenamed Kolos (Spike). This may have been based on the HASAG 30mm follow-on to the 20mm Fliegerfaust, but details are lacking. The Kolos launcher was armed with seven 30mm NRS-30 rockets compared to five on the German weapon. The NRS-30 rockets had a booster charge to eject them out of the launch tubes, followed by a sustainer motor to give them a range of 500m against helicopters and 2,000m against ground targets. The warhead could penetrate up to 10mm of armor. The weight of the system was about 14kg (31lb). The first system was built in April 1967 and underwent tests from June 1967 through May 1968. A salvo shot had a 14 percent probability-of-hit against a hovering helicopter at 500m, reduced to 4 percent against a moving helicopter. The Soviet Army had little interest in such a weapon once the Strela-2 matured, however, but some consideration was given to manufacturing the weapon for use by North Vietnamese forces, who were requesting an antiaircraft missile to deal with US helicopters. In the event, no production was authorized. It is worth noting that the People's Army of (North) Vietnam (PAVN) used the RPG-7 (*Ruchnoy Protivotankoviy Granatomyot*: hand-held antitank grenade launcher) – a more versatile solution than a dedicated unguided antiaircraft rocket – with some success against US helicopters in Vietnam (see page 49).

The Kolos was a short-lived Soviet attempt to revive the German Luftfaust concept. It used five 30mm NRS-30 unguided rockets. By the time development was complete, however, the problems associated with the Strela-2 had been solved. As a result, Kolos was canceled.

vehicle-mounted 9K31 Strela-1 (SA-9 "Gaskin") for defense of tank divisions; and the smaller, man-portable 9K32 Strela-2 (SA-7 "Grail") for the defense of motor-rifle and airborne divisions. *Strela* is the Russian word for "arrow." Both programs formally began by virtue of a government decree on August 25, 1960.

The Strela-2 was very similar to the Redeye in many respects including the basic layout. One of the differences between the two systems was that Strela-2 used a detachable 9P53 gripstock that attached to a 9P54 missile container/launch tube in a configuration similar to that of the early Redeye rather than the production version of the Redeye.

Initial unguided ballistic tests of the missile began in 1962. Strela-2 propulsion differed from the Redeye as it did not have a discrete ejection booster. Instead, the Strela-2 had a three-phase solid-rocket motor. On ignition, 34 sticks of extruded propellant burned for 0.05 seconds, ejecting the 9M32 missile from the launch tube. A powder train ignited by this process burned for 0.25 seconds to allow the missile sufficient time to clear the launch tube to avoid the gunner being injured by the exhaust from the main rocket charge. Once the main motor was ignited, the booster portion burned for 1.8 seconds followed by a 6-second sustainer burn, for a total engine burn time of about 8 seconds. A gas generator in the guidance section burned for about 8 seconds to provide air pressure for the flight-control surfaces.

Development of the guidance seeker was the single most challenging aspect of the Strela-2 program. Guided tests began in 1963 but instead of the 127 seekers that were supposed to be delivered by LOMO, only four were provided that year. In 1964, LOMO delivered 12 instead of 155, further slowing the program. Factory tests of the guided missile began in 1964 but were halted in September 1964 due to poor seeker performance. Of the 55 tests conducted through mid-May 1966, 33 failed, mainly due to seeker issues.

Maturation of the Strela-2's 9E42 seeker late in 1966 and early 1967 was successful enough that state tests began later in 1967 at the NIZAP (*Nauchno-ispaytatelniy zenitno-artilleriyskiy poligon*; anti-aircraft Research-Testing Proving Ground) near Donguz on the Orenburg steppes. The Strela-2 was formally accepted for Soviet Army use in January 1968 and production started at the Degtyaryev plant in Kovrov later in 1968. Further Soviet Army training and test launches continued over the next several years with 2,960 launches in 1968, 3,650 in 1969, and 4,000 in 1970 to overcome faults associated with the early serial-production missiles. The Degtyaryev plant was awarded the prestigious Order of the October Revolution on January 18, 1971, due to its success in mastering production of the new missiles.

The 9M32 Strela-2 missile was less sophisticated than the XFIM-92A Redeye because it relied on a first-generation uncooled lead-sulfide detector. Not only did this limit its use to tail-chase launches, but it also made the seeker vulnerable to being blinded if oriented too near the sun or by sun glint off clouds. This shortcoming was understood by the Soviet Army, and as a result, the Strela-2M upgrade program was started on September 2, 1968.

The aim of the Strela-2M program was to develop the 9E46 seeker with an electrically cooled detector similar to the early Redeye's Mod 60 detector. The Soviet detector had been developed by GOI in Leningrad. An improved rocket motor was also introduced to permit the engagement of targets up to speeds of 540km/h. The gripstock was the improved 9P53M, but this was not reverse-compatible with the earlier 9K32 system due to a different electrical interface. The two launchers can be distinguished by the location of the audio warning indicator that was on the underside of the original 9P53 Strela-2 gripstock but on the left side of the improved 9P53M Strela-2M gripstock and so closer to the gunner's ear.

A short-term solution to the fratricide problem with the Strela-2M was the introduction of a PRP. This was a small antenna fitted to the operator's helmet that detected the radio-frequency emissions of hostile aircraft. This Strela-2M is in use with the Polish People's Army in the 1970s. The yellow color of the launcher identifies it as a trainer, not a combat weapon.

THE STRELA-2M REVEALED

9K32M Strela-2M missile system

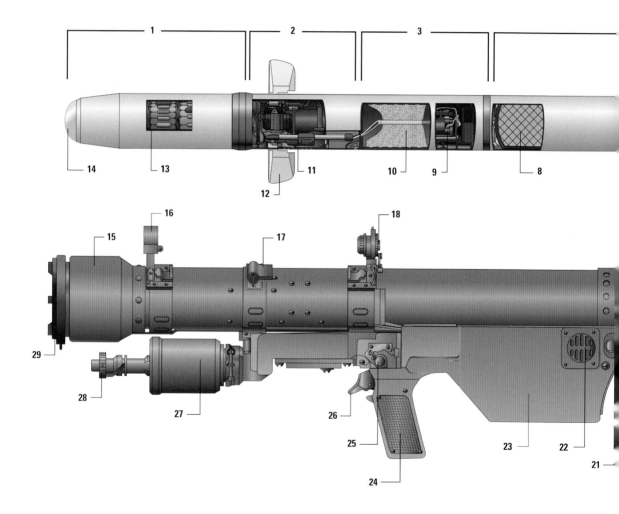

1. Seeker section
2. Guidance section
3. Warhead and fuze section
4. Rocket motor section
5. Fin and rocket exhaust section
6. Tail fins
7. Exhaust nozzle
8. Rocket fuel
9. Fuze
10. Warhead

11. Guidance motors
12. Guidance fins
13. Seeker electronics
14. Seeker lens
15. 9P54M transport/launch canister
16. Front sight
17. Front strap attachment
18. Rear sight
19. Rear strap attachment
20. Blowout disk

21. Locking pin
22. Audio warning indicator
23. 9P53M gripstock
24. Pistol grip
25. Safety switch
26. Trigger
27. 9B17 thermal battery
28. Battery actuator
29. Front cover

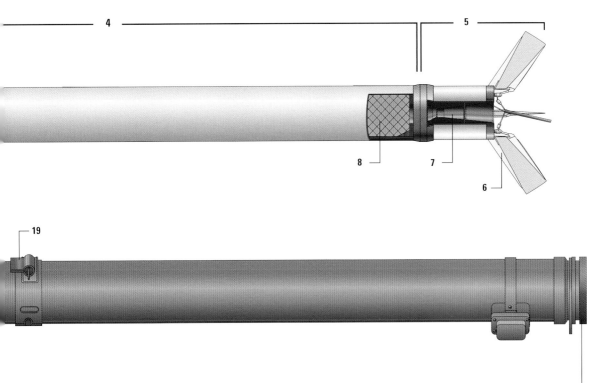

State trials of the Strela-2M began in October 1969 and were completed in February 1970. As a result of the tests, the 9K32M Strela-2M missile system with the 9M32M missile was accepted for Soviet Army use on February 16, 1970, and put into production at the Degtyaryev plant in Kovrov. The Strela-2M was roughly equivalent to the initial XFIM-43A Redeye in terms of its seeker performance and so was roughly five years behind the US system.

A further iteration of this missile, the 9M32M2 Strela-2M2, used compressed carbon dioxide for seeker cooling. Although the Strela-2M2 was not acquired by the Soviet Army due to the advent of the more advanced Strela-3 that used superior nitrogen cooling, it was supplied to export clients for licensed manufacture. The shortcomings of the Strela-2M were clearly understood by the Soviet Army, but the system offered enough improvements to justify its mass-production.

STRELA-2/-3 COMPARATIVE TECHNICAL CHARACTERISTICS			
Weapon name	Strela-2	Strela-2M	Strela-3
System designation	9K32	9K32M	9K34
Missile designation	9M32	9M32M	9M36
Gripstock designation	9P53	9P53M	9P58
Transport/launch tube	9P54	9P54M	9P59
Seeker	9E42	9E46	9E45
US DoD designation	SA-7a	SA-7b	SA-14
ASCC reporting name	"Grail"	"Grail Mod. 1"	"Gremlin"
Seeker cooling	Uncooled	Electrically cooled	Cryogenically cooled
Seeker detection	1.7–2.8µm	1.7–2.8µm	3.5–5µm
System weight	14.5kg	15kg	17kg
Missile weight	9.15kg	9.15kg	10.3kg
Warhead weight	1.17kg	1.17kg	1.1kg
Engagement range	800–3,600m	800–4,200m	500–4,500m
Engagement altitude	50–1,500m	50–2,300m	50–3,000m
Maximum speed of withdrawing target	790km/h	935km/h	1,115km/h
Maximum speed of approaching target	n/a	540km/h	935km/h
Average missile speed	1,550km/h	1,550km/h	1,440km/h
Probability of hitting withdrawing aircraft	19–25%	22–25%	31–33%
Accepted for service	1968	1970	1974

The final iteration of the Strela-2 family was codenamed Strela-3. The 9K34 Strela-3 program was started on September 2, 1968, at the same time as the Strela-2M upgrade effort. The main aim of the program was to develop the new 9E45 cryogenically cooled seeker. Unlike the previous seekers developed in Leningrad, this program was supervised by the Arsenal Plant in Kiev, one of the Soviet Union's largest missile seeker manufacturers. Development was undertaken under chief designer I.K. Polosin of the Central Design Bureau Tochnost. This design bureau was part of the Progress plant in Nizhyn that manufactured the seekers in cooperation with the nearby Arsenal Plant. The Strela-3 can be

AN EXPORT GIANT

The 9K32M Strela-2M became the most significant first-generation MANPADS due to its widespread proliferation and extensive combat use. Over 310,000 9M32 Strela-2 and 9M32M Strela-2M missiles were manufactured in the Soviet Union from 1968 through 1983 of which about 230,000 were acquired by the Soviet armed forces and about 80,000 for export. The figures below are based on US intelligence estimates because the Soviet Union never published any data on the scale of missile production. The Soviet Union encouraged Warsaw Pact countries to begin licensed production. Poland manufactured 7,000 Strela-2M under license while Romania manufactured 2,280. Czechoslovakia manufactured the Strela-2M until 1977, with ZVS as the prime contractor and Konštrukta Trenčín providing key subassemblies. The Soviet Union also sold licensed-production rights to allied countries including Egypt.

Other countries obtained the technology through subterfuge. Starting in 1975, China reverse-engineered the Strela-2 as the HN-5 (Hongying-5; "Red Tassel-5"), but there are disputes about the origins of these missiles. Russian accounts claim the program was based on two Strela-2 missiles taken from shipments to Vietnam that passed through China; other accounts suggest an Egyptian source. The reverse-engineering program was concentrated in Liaoning province with Factory No. 119 responsible for missile final assembly. Test launches began in March 1975, but further testing in 1980 revealed problems with premature warhead detonation that delayed design certification until April 1985. While these efforts were taking place, a parallel-manufacturing program was undertaken in the Shanghai region, the traditional center of Chinese surface-to-air missile development. Tests conducted in 1981 showed that the Shanghai missile, designated HN-5A, had poor reliability. Design certification of the HQ-5A (Hongqi-5; "Red Flag-5") was completed in November 1986. This missile had a larger warhead and improved IRCM features compared to the HQ-5, and probably was based on examples of the Strela-2M vs. the earlier Soviet Strela-2. Pakistan license-manufactured the HQ-5A as the Anza Mk I.

INTERNATIONAL LICENSED PRODUCTION OF STRELA-2/-2M MISSILE		
Plant	**Location**	**Local name**
Vazovski Machinostroitelni Zavodi	Sopot, Bulgaria	Strela-2M
Závody všeobecného strojárstva	Dubnica nad Vahom, Czechoslovakia	Strela-2M
Zakłady Metalowe Mesko	Skarżysko-Kamienna, Poland	Strzała-2M
Arsenalul Armatei and Electroomecanica	Bucharest and Ploesti, Romania	CA 94, CA 94M
Krušik Valjevo	Valievo, Yugoslavia	Strela-2
Sakr Factory No. 81	Heliopolis, Egypt	Ayn-al-Saqr, Sakr-Eye, Falcon's-eye
Shahid Shah Abhady Industrial Complex	Tehran, Iran	Sahand
4th Industrial Bureau	Pyongyang, North Korea	Hwasung-Chong

distinguished from the earlier types by its new 9P51 BCU on the front of the 9P58 gripstock. Instead of the cylindrical battery of the previous types, the Strela-3 had a two-part BCU with a cylindrical battery on the front connected to a ball-shaped vessel for the pressurized nitrogen gas.

The Strela-3 was comparable to the standard production version of the Redeye, the FIM-43C. As a result, it had a limited capability to engage enemy aircraft from the front and sides, but it was still most effective when used against jet aircraft targeted from the rear. The Strela-3 was

The 9K34 Strela-3 system was essentially similar to the Strela-2M except for its new cooled seeker. It can be distinguished from the previous Strela-2 family by its distinctive ball-shaped 9P51 BCU on the front of the 9P58 gripstock. (Author)

accepted for service use in 1974 and was manufactured at the Degtyaryev plant in Kovrov. Strela-2M production continued alongside that of the Strela-3 until 1983 because the superior Strela-3 had not been cleared for export and was initially manufactured exclusively for the Soviet Army. Total Strela-3 production was around 56,000 9M36 missiles through 1988 with very modest export production.

BRITISH MANPADS: BLOWPIPE, JAVELIN, STARBURST, AND STARSTREAK

Short Brothers in Northern Ireland began exploring a MANPADS in 1965–66 as a private company venture. The firm had already conducted experiments in the early 1960s, dubbed SX-A5, to convert the Malkara wire-guided antitank missile into a short-range surface-to-air missile by substituting radio-command guidance. This led to the Green Light prototype, which eventually emerged as the ship-based Seacat and land-based Tigercat air defense missile systems. In 1965–66, Short Brothers began a private venture to determine whether this guidance approach could be scaled down to a man-portable system. This was offered to the British Army, with development funding beginning in 1967 and initial test firings in 1968. Unlike the Redeye and Strela-2, the Blowpipe used MCLOS guidance via a radio link between the launcher and the missile.

In the initial stage of flight, the Blowpipe missile was automatically gathered into the operator's line of sight by tracking IR flares mounted on the rear of the missile. Guidance continued with the gunner tracking the target and steering the missile via a thumb control, with instructions being transmitted to the missile by a radio link. Control was via four canard guidance fins, with one pair providing roll control and the other pair providing yaw and pitch.

The Blowpipe missile was slightly heavier than the Redeye and Strela-2, at 11kg (24lb). The greatest difference was the heavier overall system weight, 21.4kg (47lb), because the launcher included an aiming unit, tracker, and battery power supply. Aside from the large tracker/aiming unit, the Blowpipe system was also very distinctive due to the enlarged cylinder at the front of the launcher. This was designed to accommodate the four canard guidance fins because unlike the Redeye and Strela-2, they were not folded inward prior to launch but remained in the extended position.

The Blowpipe launcher configuration was substantially different from other MANPADS of the period due to the use of a large launcher control to guide the missile. The larger diameter of the forward section of the launch tube was due to the four canard guidance fins of the missile which did not fold into the missile body as on most other MANPADS. (US DoD)

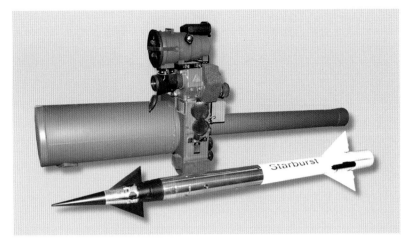

The Javelin S15/Starburst closely resembled the Blowpipe because the major changes were internal to permit SACLOS guidance. This particular example is fitted with a Pilkington night sight to permit night firing. (Author)

Final British Army trials of the Blowpipe system were scheduled to end in April 1970, but dragged on through 1975. A low-rate production contract was awarded to Short Brothers in September 1972 based on an initial British Army order for 285 systems. The Blowpipe entered operational service in 1975 following the final certification tests. The British Army order was followed shortly after by a Canadian order for 100 systems in June 1973. Export customers later included Nigeria, Oman, and Thailand.

As was the case with many other MANPADS in the 1970s, the Blowpipe was upgraded with a Cossor IFF system based on a British Army order in May 1977. The weight of the Blowpipe system encouraged the development of the Lightweight Multiple Launcher (LML), a pedestal launch post that could accommodate three Blowpipe missiles and a tracker. The LML was acquired for missions in which the man-portability feature was not essential, such as site defense.

The Blowpipe's combat debut during the 1982 Falklands War was very disappointing (see page 55). It suffered from the same inherent problems as most MCLOS guidance systems in that manual control of the missile during the short launch–intercept sequence proved very difficult in the chaos of actual combat. There had been a general trend away from MCLOS to SACLOS guidance systems with antitank missiles in the 1960s, made possible by improvements in digital microprocessors. Not surprisingly, this was the choice to upgrade the Blowpipe. SACLOS guidance systems lessened the workload of the gunner by using a more sophisticated tracker that automatically followed the target and missile. The gunner merely had to keep the crosshairs of the aiming unit on the target, and the tracker then correlated the two objects and automatically sent signals to the missile via the usual radio-command channel to intercept the target.

The SACLOS version of the Blowpipe entered production in 1984 as the Javelin missile. By this stage, the British Ministry of Defence had released a General Staff Requirement (GSR 3979) for a new air defense system, which would eventually emerge a decade later as the Starstreak. In the interim, the Javelin was upgraded with a laser-command link to

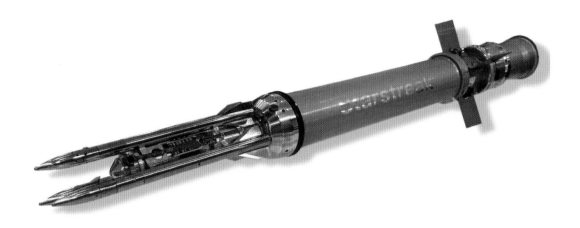

The Starstreak missile is unique among MANPADS for its unusual warhead, consisting of three laser-guided "hittile" darts rather than a conventional high-explosive warhead. (Author)

replace the existing radio link under the designation Javelin S15. This system was renamed as the Starburst in the late 1980s when it was cleared for export. The laser-command link, less susceptible to interference or deliberate jamming than a radio link, stemmed from work on the later Starstreak. The Starburst was viewed as an intermediate solution until the Starstreak was ready. The Starburst entered production in 1988 for the British Army and remained in production for export even after the Starstreak entered production in 1997 due to export restrictions on the latter system. Export of the Starburst/Javelin S15 began in 1992 with an initial order from Canada. The Starburst was deployed during Operation *Telic* in Iraq with 40 Regiment, Royal Artillery.

The new GSR 3979 air defense missile program started in January 1985 with the award of two competitive project-definition contracts for Short Brothers' Starstreak HVM (High Velocity Missile) and British Aerospace Dynamics' Thunderbolt. In June 1986, the British Ministry of Defence selected the Starstreak. The system used SACLOS guidance with a laser for the command link between the launcher and missile. The Starstreak has a unique warhead that includes three small darts, dubbed "hittiles," instead of the usual unitary high-explosive warhead. These darts separate from the core missile after the main rocket motor burns out. Each dart contains its own guidance and control circuitry, powered by a thermal battery, using a laser from the launcher for the terminal guidance.

The Starstreak was designed from the outset for three configurations. The Armoured Starstreak on the Stormer HVM armored vehicle was fielded first, followed by the LML pedestal, and finally the man-portable version. There were significant program delays due to the novelty of the warhead. The first successful shoulder-launch test of the missile was carried out in mid-1988. Production of the Starstreak system began in 1993, and it was initially deployed in 1994. The Starstreak system was finally cleared for export in 1998 and won contracts from South Africa, Indonesia, Malaysia, and Thailand. After the sale of the company and several reorganizations, Short Brothers became Thales Air Defence Limited in 2001.

THE NEXT GENERATION: STINGER

Development of a next-generation Redeye II in 1965 focused on a substantially enhanced guidance system. The first generation of IR seekers had a rotating rectangular field of view and single detector element. Aside from their susceptibility to IRCM, this scanning technique led to increasing inaccuracy as the missile arrived near the target. The second generation of IR seekers switched from rotating the reticle to rotating the optics with a conical scanning technique that eliminated these inaccuracies. The final phase of the Advanced Sensor Development program was completed at General Dynamics' Electro-Dynamics Division (formerly Convair) in December 1970. In February 1971, Redstone Arsenal evaluated the General Dynamics concept against six other alternatives and decided to proceed with the development of the XFIM-92A based on the new reticle-scan seeker, which was sensitive to IR energy in the 4.1–4.4μm band. For terminal flight control, the reticle-scan seeker used target adaptive guidance, which biases missile trajectory toward vulnerable portions of the target airframe to assure maximum lethality.

In March 1972, Redeye II was renamed as Stinger. Several other features were sought in the new missile system, including resistance to countermeasures, and the incorporation of an IFF system. Guided flight

The Starstreak is most commonly used from a Lightweight Multiple Launcher (LML) pedestal launch post or from a Stormer HVM armored vehicle. There is also a MANPADS version, shown here during a US/UK Combined Joint Operational Access Exercise at Fort Bragg, North Carolina, in 2015. The blue coloring of the missile launch tube identifies this as a training round. (Capt. Joseph Bush, 82nd Airborne Artillery, Division Public Affairs)

tests of the Stinger began in November 1973, and in February 1975 a guided test missile first scored a direct hit against a target traveling at 400 knots (460mph). Problems during initial guided tests in 1974 led the US Army to fund a back-up missile, Ford Aerospace's Saber laser-guided weapon. The Saber was very similar in design and layout to the British Starburst and development continued until 1977, by which time it was clear that the Stinger would meet its objectives.

The FIM-92A Stinger was type-classified as standard by the US Army in November 1977, clearing the way for serial production. This became the baseline version of the Stinger and was manufactured from 1977 through 1987 with a total of 15,669 produced. The FIM-92A Stinger reached initial operational capability in February 1981 with US Army troops based in Europe.

In contrast to the Redeye, the Stinger had a detachable gripstock. After a missile was fired, the transport/launch tube was removed and a new round clipped to the gripstock. The Stinger gripstock could be fitted with an AN/PXX-1 IFF antenna on the forward right side of the gripstock. The IFF antenna connected to an IFF interrogator box, worn on the gunner's belt. The system notified the gunner of a Mode 4 (friend) or Mode 3 (possible friend) by an audible signal.

In 1976, work began on a third-generation seeker called POST (Passive Optical Seeker Technique). This seeker used rosette-scan optical processing that was sensitive in both IR and UV bands to circumvent common IRCM. The POST seeker was approved for production in June 1983 as the FIM-92B Stinger-POST and first deliveries took place in September 1986. This was the most short-lived of the different Stinger versions because in November 1985, the US Army decided to switch to the upgraded Stinger-RMP.

In 1984, the US Army began development of a reprogrammable microprocessor (RMP) that allowed threat updates to be incorporated by changing software in the Stinger gripstock rather than hardware. The RMP feature was added to the existing POST improvements as well as additional counter-IRCM rejection features. Production of the FIM-92C Stinger-RMP began in November 1987 and replaced both earlier models

The FIM-92 Stinger gripstock is similar in appearance to that of the earlier Redeye gripstock with the important exception of the prominent IFF antenna mounted on the right side. (US Army)

of the missile. It was first deployed with the US Army in July 1989 and was followed by the FIM-92D RMP that introduced further counter-IRCM features. Production of new Stinger missiles for the US armed forces ended in January 2005 after about 50,700 of all Stinger types had been manufactured.

The Stinger-RMP product improvement program (PIP) began in 1992 to upgrade the older FIM-92A/-92B systems with the FIM-92C/-92D RMP features, as well as improved counter-IRCM software and a new roll sensor. The PIP upgrade was adopted as the FIM-92E Stinger Block I and much of the remaining early Stinger inventory was upgraded to the Block I standard in the late 1990s. Other recent upgrade programs have been aimed at keeping the Stinger inventory viable. The FIM-92J replaced aging components to extend the Stinger's service life by an additional ten years along with a new proximity fuze better able to deal with small targets such as drones.

The West German Bundeswehr conducted comparative trials of the Stinger, Blowpipe, and the Swedish RBS 70 in 1979 and selected the Stinger. This was the first step in the creation of the NATO Stinger Project Group including West Germany, the Netherlands, Norway, Greece, Turkey, and Italy to license-manufacture the Stinger in Europe. Dornier was selected as the European prime contractor and a production contract was signed in April 1989 for the license-manufacture of the improved

A FIM-92 Stinger being launched by a US Marine Corps team from Battery A, 2nd Low Altitude Air Defense Battalion, during Exercise *Arctic Edge* at Fort Greely, Alaska, on March 15, 2018. The booster stage can be seen near the launch tube after it has separated from the missile itself. (US Marine Corps, Lance Cpl. Cody Ohira)

THE STINGER EXPOSED

FIM-92 Stinger

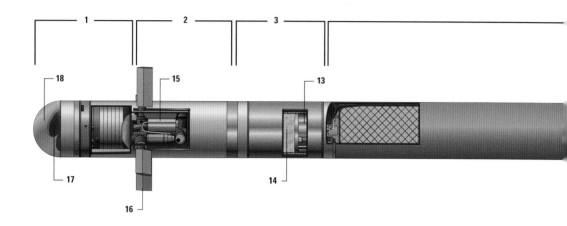

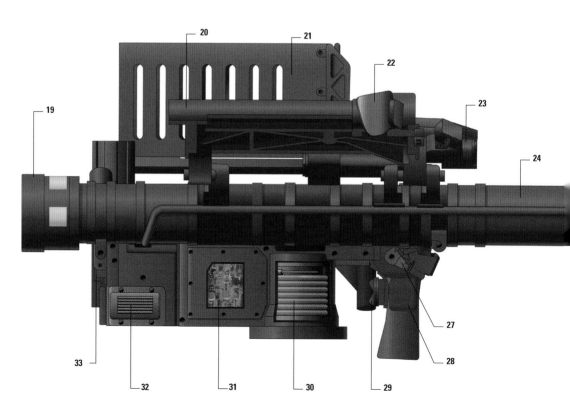

1. Seeker section
2. Guidance section
3. Warhead and fuze section
4. Rocket motor section
5. Fin and rocket-exhaust section
6. Eject motor
7. Igniter
8. Exhaust nozzles
9. Propellant
10. Tail fin
11. Ignition interlock
12. Rocket fuel
13. Impact fuze section
14. Warhead explosive
15. Advanced proportional
    navigation guidance
16. Rolling airframe canard control fins
17. Seeker rosette scan
18. Seeker head dome
19. Front cover
20. Range finder
21. IFF antenna
22. Eye shield
23. Rear sight
24. Launch tube
25. Squib leads
26. Blowout disk
27. Safety and actuator
28. Gripstock assembly
29. Trigger
30. Battery/coolant unit
31. Control circuit board
32. Uncaging switch
33. Latch mechanism

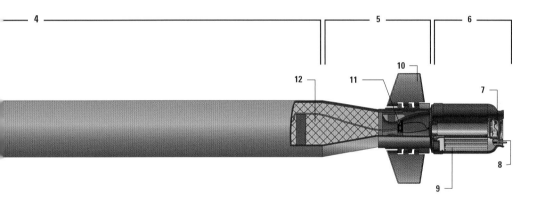

MANPADS are too expensive to launch during day-to-day training, so most systems have dedicated training versions. This is the Stinger Tracking Head Trainer that contains an active seeker to accustom the gunner to the launch process. This version can be most easily identified by the Performance Indicator Assembly, the box mounted on the rear of the launch tube. The blue markings on the system also indicate a trainer; war rounds have yellow markings. (Author)

FIM-92C Stinger-RMP. Switzerland also manufactured FIM-92C Stinger without the RMP features for local requirements based on a February 1988 agreement. The Stinger has been adopted by over 20 countries with the FIM-92F being the current production standard for export sales.

In the late 1990s, the US Army began developing the Advanced Stinger Block II that employed a fifth-generation focal-plane array seeker. This was a true imaging seeker that was virtually invulnerable to optical countermeasures. The program was expected to take eight years and cost $630 million; but because the Stinger had never been fired in anger by US forces, there was some question as to whether such a program was warranted. In 1999, the program was canceled during deliberations over the Fiscal Year 2001 US defense budget. This effort was not revived for two decades until the US Army began a Stinger follow-on program in 2021.

THE SOVIET RESPONSE: IGLA AND VERBA

In the early 1970s, development of a successor to the Strela-2 series began at the KBM design bureau in Kolomna under the codename *Igla* (Needle) and based on a government authorization of February 12, 1971. As in the case of the Stinger, the main aim was to develop an improved guidance system to permit all-aspect engagement of aircraft targets. There was also a desire to extend the effective range of the missile because NATO attack helicopters and strike aircraft were being fitted with standoff missiles that exceeded the maximum range of the earlier Strela-2 and Strela-3 family of missiles. The Soviet Army subsequently added a requirement for an IFF system based on the lessons from early Egyptian use of the Strela-2, which revealed an alarming tendency of MANPADS gunners to fire on friendly aircraft due to poor aircraft-recognition skills. The new Igla system also included a more sophisticated infrastructure of cueing aids to warn the Igla gunner of the approach of enemy aircraft.

The 9E410 seeker was developed by LOMO. The development effort encountered severe problems with the seeker due to attempts to develop exotic guidance technology such as staring arrays, image sensors consisting of an array of light-sensing detectors at the focal plane of a lens. As a result, the Igla program was substantially reorganized and restarted by a second government decree on January 18, 1974.

Owing to the delays this reorganization and restart imposed, a parallel program was started in 1978, the Igla-1. The 9M313 missile shared most of the new components of the Igla's 9M39 missile, but employed the less sophisticated 9E418 one-channel cooled IR seeker developed under I.K. Polosin of the Central Design Bureau Tochnost, the bureau that had developed the previous Strela-3 seeker. An intriguing difference between the Soviet and US seekers was that the Soviet seekers employed an aerodynamic spike (aerospike) on the nose of the missile to improve the speed of the missile. On the Igla-1 missile, a cone was fitted in front of the seeker, suspended on a wire tripod. The Igla-1 missile corresponded roughly to the FIM-92A Stinger missile.

The 9K310 Igla-1 was formally adopted by the Soviet Army in 1981, two years earlier than the more sophisticated 9K38 Igla. The Igla-1 systems manufactured for the Soviet Army could be fit with the 1L14 IFF system, called an NRZ (*nazemniy radiozaprochik*; ground radio interrogator). In the event of a positive response from a friendly aircraft, the NRZ prevented the missile from being fired. The NRZ antenna was contained in a lightweight, horseshoe-shaped foam-plastic cover on the front of the missile launch tube and used in conjunction with the upgraded 9P519-1 gripstock with the added 31AR interrogator module underneath.

The 9K310 Igla-1 somewhat resembles the previous 9K34 Strela-3, with a similar BCU positioned at the front of the gripstock. However, it can be distinguished by the different shape of the fluted front cover of the transport/launch tube that was elongated to shield the aerospike on the nose of the 9M313 Igla-1 missile. (Author)

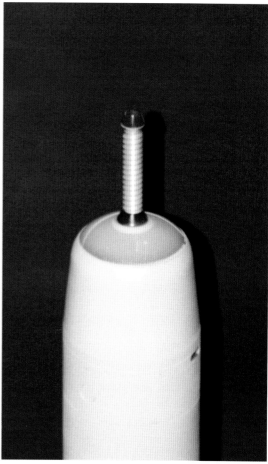

One of the interesting innovations on the Igla and Igla-1 missiles was the use of an aerospike on the front of the seeker to improve missile speed. On the initial 9M313 Igla-1 missile (left), the aerospike consisted of a cone suspended on a small tripod over the seeker head. On the later 9M39 Igla missile (right), the aerospike was replaced by a simple probe. While it might seem this would interfere with the seeker, it was located over the central reflector of the Cassegrain telescope, and so did not block the collection of the main mirror. (Author)

Owing to the delay in the development of the 9E410 seeker, the Igla system was not accepted for Soviet Army use until 1983. The new seeker operated in both the IR and UV bands, which made it far less susceptible to IRCM such as flares because the UV detector was not spoofed by the flares. It was also effective against the early lamp-style IR jammers, though not the later shutter-type. The 9E410 seeker was about twice as sensitive as the 9E418 seeker on the Igla-1. The military authorities in Finland, which operated both the Igla in its army and the French Mistral in its navy, felt that the Igla had a superior seeker to the Mistral.

As in the case of the Igla-1, the Igla system used by the Soviet Army could be fitted with the 1L14 IFF system. An upgraded IFF antenna was also developed, the I-7R, which clipped on to the 9P516 gripstock above the launch tube. The full configuration with the I-7R antenna is seldom encountered, however, and it is not clear if it entered serial production.

The Igla MANPADS was accompanied by increasingly sophisticated targeting aids. The first of these was the 1L10 PEP (*perenosnoy elektronniy planshet*; portable electronic tablet); it included a small screen that displayed target information such as incoming hostile aircraft, based on data fed from the regimental air defense network. This was similar to the US Army's Target Alert Data Display Set (TADDS) for the Redeye. The

The Igla launcher could be fitted with an NRZ IFF antenna on the front of the launch tube. This consisted of a horseshoe-shaped block of plastic foam that protected the IFF antenna embedded inside. This particular 9K38 Igla does not have its BCU fitted. (Author)

1L10 PEP was followed in the late 1980s by the improved 1L15-1 PEP that increased the reporting area for the device.

A number of Igla-1 derivatives were developed in Russia, but were apparently produced in small numbers, if at all. The Igla-1D was intended for Soviet airborne forces and used a two-part missile/transporter/launcher to reduce the overall length of the system when parachuting. The Igla-1N used a larger warhead. The Ukrainian firm Immersion has offered an upgraded Igla for export under the designation PZRK 336-24. This uses an improved seeker developed by the Central Design Bureau Tochnost, which had been responsible for several of the Soviet-era MANPADS seekers.

Production of the Igla and Igla-1 in the Soviet Union totaled about 74,000 from 1978 to 1988; data for later manufacture is lacking. The Soviet Union was by far the largest exporter of MANPADS in the final decades of the Cold War and in the decades since. From 1975 to 2015, the Soviet Union and then Russia exported about 105,000 MANPADS, the Igla family accounting for about 23,000 of these. By way of comparison, total US Stinger exports, not counting European coproduction, were fewer than 11,000 missiles, only one-tenth of the Soviet total.

Development of a next-generation Soviet MANPADS, codenamed *Verba* (Willow), began in the early 1990s. This was intended to incorporate an advanced multiwavelength seeker that offered multispectral capabilities to defeat any existing countermeasures. This program suffered from serious delays due to a number of factors, however, starting with a severe decline in research funding after the breakup of the Soviet Union in 1991. The breakup also affected cooperation between Russia and the former Soviet republics, especially Ukraine. The Tochnost/Progress/Arsenal conglomerate in Ukraine had been a major developer of MANPADS seekers, and it was now cut off. As a result, the Verba program was forced to rely on LOMO for the new seeker. The program was severely constrained by a lack of new funding and a hemorrhage of skilled engineers at many defense research facilities due to lack of pay.

IGLA AND VERBA COMPARATIVE TECHNICAL CHARACTERISTICS				
System name	**Igla**	**Igla-1**	**Igla-S**	**Verba**
System designation	9K38	9K310	9K338	9K333
Missile designation	9M39	9M313	9M342	9M336
Gripstock designation	9P516	9P519	9P522	9P521
Seeker	9E410	9E418	9E435	
ASCC/DoD designation	SA-18	SA-16	SA-24	SA-29
NATO reporting name	"Grouse"	"Gimlet"	"Grinch"	"Gizmo"
System weight	18.4kg	17.9kg	19kg	17.25kg
Missile weight	10.8kg	10.8kg	11.3kg	12kg
Warhead weight	1.17kg	1.17kg	2.5kg	2.5kg
Engagement range	500–5,200m	500–5,200m	500–6,000m	500–6,400m
Engagement altitude	10–2,500m	10–2,500m	10–3,500m	10–4,500m
Maximum speed of withdrawing target	320km/h	320km/h	320km/h	320km/h
Maximum speed of approaching target	360km/h	360km/h	400km/h	500km/h
Average missile speed	570km/h	570km/h	570km/h	570km/h
Probability of hitting withdrawing aircraft	45–63%	38–59%	50–75%	60–90%
Accepted for service	1983	1981	2002	2014

The standard version of the Igla since 2002 is the 9K338 Igla-S. This has a larger warhead than earlier Igla variants and is fitted with optical proximity fuzes that can be seen around the periphery of the 9M342 missile behind the front steering fins. (Author)

The new MANPADS program was restarted in the late 1990s, adopting a two-track approach similar to the previous Igla/Igla-1 programs. In the short-term, the new 9K338 Igla-S (Super) adopted some elements from the Verba program including the addition of a proximity fuze, a new and more efficient rocket motor, and the enlarged 9N330 2.5kg warhead. However, the Igla-S used the LOMO 9E435 seeker instead of the more challenging Verba seeker. State tests of the Igla-S were completed in December 2001 and the system was accepted for service in 2002. It was put into production at the Degtyaryev plant in Kovrov and became the principal export version of the Igla missile in the new century.

The Verba program dragged on at a slow pace due to lack of funding. The Verba seeker operates in three wavelengths, near-IR, mid-IR, and UV, in order to counteract contemporary aircraft countermeasures. Initial test launches conducted by the KBM missile design bureau began

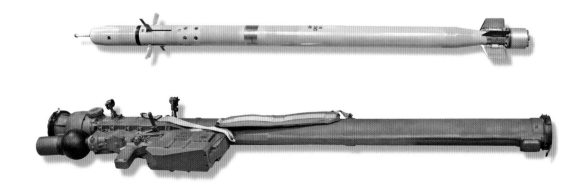

THE IGLA ABROAD

The 9K310-1 Igla-1M system was developed in the early 1980s specifically for manufacture by Warsaw Pact countries, notably Poland and Bulgaria. It used the 9P519-2 gripstock that initially lacked the 1L14 IFF system. Likewise, the international export version of the system, the 9K310E Igla-1E (E for *eksportny*) used the 9P519E gripstock without the IFF system. As part of a Warsaw Pact standardization program, the Soviet government encouraged the Polish and Bulgarian production efforts in order to supply the other Warsaw Pact armies with MANPADS.

In 1989–90, Poland negotiated the licensed production of the Igla-1E from the Soviet Union. This involved production of many of the subsystems including the seeker, but the purchase of some subcomponents directly from the Soviet Union. The collapse of the Soviet Union in 1991 complicated this process and in 1992 the Polish government decided to "Polonize" the missile and to manufacture it indigenously as the Grom (Thunder). At the same time, many of the subsystems of the missile and launcher were upgraded. By 2000, the Grom was based entirely on Polish components. It has been exported to a number of countries including Georgia and Lithuania. Once production was underway, the Grom-M development program was initiated for a thorough upgrade of the system with a new seeker and other improved features. Once development was complete, the new missile system was renamed Piorun (Lightning). The Polish government placed its first order for the Piorun missile system in December 2016, with the first deliveries being made in 2019.

Macedonian soldiers, supervised by a Slovenian military instructor, use an Igla during a live-fire exercise in September 2008. (ROBERT ATANASOVSKI/AFP via Getty Images)

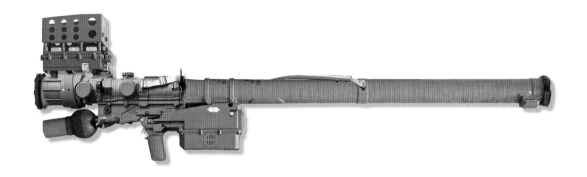

The 9K333 Verba is the latest iteration of the Igla family. It can be fitted with the 1L229V IFF antenna on the front of the transport/launch tube as shown here.

in 2007, anticipating a start of production in 2008; but the war between Russia and Georgia in 2008 revealed a problem in conflicts in the "near-abroad." Because both sides used Soviet-era aircraft, the IFF equipment could not distinguish between Georgian and Russian Su-25 strike aircraft. As a result, the Verba introduced a substantially modernized IFF system, the 1L229V. This consists of two plate antennas clipped to the upper front of the launch tube, as well as new electronics in the gripstock. The new antennae are roughly similar in appearance to the type used on the US Army Stinger, though centrally mounted rather than offset to the right. State trials of the Verba system were first postponed to 2009–10 and finally began in the summer of 2011 at the Yeysk proving ground. The tests were also behind schedule, not concluding until 2014. The first production 9K333 Verba systems were delivered to the Russian Army's 98th Airborne Division in May 2014.

The Verba continues the trend in MANPADS to integrate these missiles into regimental and divisional air defense nets. The Verba was issued in brigade sets that included the 9S933 Barnaul command-and-control system. This included specialized command vehicles at brigade and regimental level to direct the dispersed MANPADS squads. The Verba system also introduced the new 1L122 Garmon forward-alert radar, used by the Barnaul system to detect and track hostile air threats. Individual MANPADS can be fitted with the 9S935 cueing aid that includes a digital display for the gunner, alerting him of up to four targets as well as range data and other information. This cueing aid is linked to the subunit commander via a small R-168-0.5 radio carried by the gunner. The Verba also included a comprehensive suite of other support equipment including clip-on night-sights, training systems, and mobile maintenance systems. MANPADS are no longer isolated, autonomous weapons, but are part of a larger chain of the air defense network.

Russia started development of a new-generation MANPADS in 2017 under the codename *Metka* (Marker).

EUROPEAN OPTIONS: RBS 70, MISTRAL, AND SUNGUR

Some European armies were not convinced of the need for true portability and preferred a heavier but more capable system. Bofors in Sweden examined several options for a replacement for the venerable 40mm antiaircraft gun. The Robotsystem 70 program formally began in 1969 based on a Swedish Army requirement. Unlike the Redeye and Strela-2, the RBS 70 relied on a laser beam-rider guidance system to minimize its susceptibility to IRCM. Although the RBS 70 missile was not especially large, the firing unit with associated guidance controls brought the weight of a combat-ready system to 87kg (190lb). A launch post and a single missile could be carried short distances by a three-man crew, but the RBS 70 was generally deployed with a vehicle. The missile had an unusual configuration compared to the Redeye/Stinger approach because the missile guidance elements were housed in the tail of the missile facing back to the launcher, with the rocket motor in the center. Development of the RBS 70 was completed in 1975 and it went into service with the

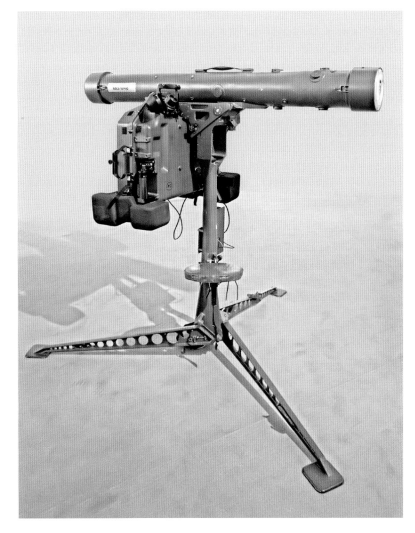

The Swedish RBS 70 was one of the heavier European alternatives to the more common IR-guided MANPADS. The configuration resembles the British Blowpipe, but the system uses a laser-command link and a launch post to accommodate the system's heavier weight. This is an example of the RBS 70NG (New Generation) introduced in 2011. (Author)

39

Swedish Army as well as several export clients. It was followed a few years later by the RBS 70 Mk 1 that featured an upgraded laser receiver to increase the target engagement area. The RBS 70 Mk 2 featured a larger warhead and increased range; it entered production in 1989.

In 1983, Bofors received a Swedish Army contract to develop the RBS 70M (*Morker*: night), later called the RBS 90. This version moved away from the man-portable category because the launcher is remotely controlled from the Bv 206 fire-control vehicle. The RBS 90 was intended to supplement rather than replace the RBS 70 for certain missions. In the 1990s, Bofors developed the improved Bolide missile for the RBS 90, which offered higher speed. It eventually replaced the RBS 70 Mk 2 missile for all customers after 2002.

Like many small air defense missile systems, the RBS 70 was adapted for a variety of land and naval launchers. The RBS 70NG (New Generation) was introduced in 2011. The new launcher has an integrated high-resolution thermal imager for night launches, cueing aids to improve reaction time and target acquisition, an auto-tracker to aid the gunner during the engagement, and built-in video recording for after-action review. The RBS 70 in its various iterations was adopted by 17 countries, with about 18,000 missiles manufactured through 2020.

Another heavier, pedestal-mounted missile system was the French Mistral. In the late 1970s, the French Army established its criteria for a SATCP (*Sol-air à très courte portée*: Surface-to-Air Very Short Range) missile system. In 1980, the Matra Mistral design was selected to fulfill these requirements. The French Army wanted a larger warhead with greater lethality than the Stinger/Igla approach, and was willing to sacrifice portability. The Mistral missile in its launch container weighed 24kg (53lb), nearly double the weight of the Stinger and Igla. The main attraction of the Mistral was its substantially larger warhead – at 2.9kg, it was nearly triple the weight of the Stinger (1kg) and Igla (1.2kg) warheads – but the Mistral had some other significant differences including an IR seeker that was based on four detectors arranged in a cross pattern rather than the single detector on other MANPADS. Owing to its weight, the Mistral was deployed on a pedestal launch station that weighed 62kg (137lb) when combat ready.

Development and testing was completed in 1988 and the Mistral was deployed by the French Army and Air Force using pedestal launchers, and in the French Navy using specialized launchers for shipboard use such as the Sadral. Matra developed other launcher options for the Mistral

including an air-to-air version for helicopters, and an assortment of vehicle and ship mountings. By 2000/01, total deliveries of the Mistral to the French Army totaled 125 firing posts and 1,230 missiles with a further 1,090 missiles for the French Air Force.

In 2000, Matra began development of the Mistral 2 missile with improved maneuverability, greater speed, and an expanded firing envelope. This was aimed primarily at the export market; ultimately, export orders outnumbered domestic orders by more than three-to-one. In 2008, France began an upgrade program entitled Mistral RMV (*Rénovation à mid-vie*: Midlife upgrade) for its remaining inventory. Improvements developed for this version were incorporated into the new Mistral 3 in 2014. Total Mistral production through 2020 was around 18,000 missiles of which about 4,500 were for the French armed forces and the rest for export.

Turkey decided to locally develop a replacement for its imported Stinger missiles under a contract signed with the Roketsan firm on September 10, 2013. The new missile was codenamed *Sungur* (Falcon) and test launches began in 2020. The Sungur employs an imaging-IR red seeker, greatly reducing the missile's vulnerability to IRCM. The Sungur appears to be significantly heavier than the Stinger, and so far, it is being deployed on vehicle launchers, though a gripstock version has been mentioned.

The Mistral missile was significantly larger and heavier than other MANPADS of the period to accommodate a heavier warhead. It also used a different type of IR guidance, having four detectors arranged in a cross pattern under a distinctive pyramidal seeker cover. This is an example of the Mistral 2, introduced in 1997. (Author)

CHINESE MANPADS

After copying and manufacturing the Soviet Strela-2 as the HN-5, China later undertook development of at least two more generations of improved MANPADS. Design histories of these missiles are not available, so this account relies on the dates of their debuts at various international arms shows to provide some idea of their chronology.

The best-known family of Chinese MANPADS are designated QW for *Qian Wei* (Advance Guard). They have been developed by research institutes of the 5th Academy of the Ministry of Defense, now called the China Aerospace Science and Industry Corporation (CASIC), and they are manufactured by Factory 119 in Shenyang (Shenyang Hangtian Xinle Ltd.). The public face of these systems is the China National Precision Machinery Import–Export Corporation (CPMIEC), which handles their international export. China was able to exploit foreign MANPADS technology to upgrade its own designs. Iran provided China with Stingers

Pakistan's MANPADS have been dependent on technology transfers from China. The Anza Mk II, shown here, is a close derivative of the Chinese QW-1, while the earlier Anza Mk I was based on the HN-5. (Author)

The QW-18 is the Chinese equivalent of the Russian Igla. Although some features such as the gripstock resemble to those of the Strela-2M, distinctly Chinese features appeared such as the BCU. (Author)

obtained in Afghanistan. China also managed to obtain Stinger and Igla missiles from the fighting in Angola via Zaire.

The QW-1, sometimes marketed as the Vanguard, was a follow-on to the HN-5A, but with a cooled seeker and a new BCU. It was unveiled internationally in 1994 at the Farnborough Air Show. Licensed derivatives include the Pakistani Anza Mk II and Iranian Misagh-1. The follow-on QW-2 appears to have been based on technology from the Russian Strela-3 missile and was unveiled in 1998 at the Farnborough Air Show. This system incorporates a BCU that is nearly identical to the Soviet design. The QW-11, unveiled in 2002 at Airshow China in Zhuhai, was a derivative of the QW-2 that added a laser proximity fuze, but it appears to have been short-lived. The QW-18 debuted at the 2007 International

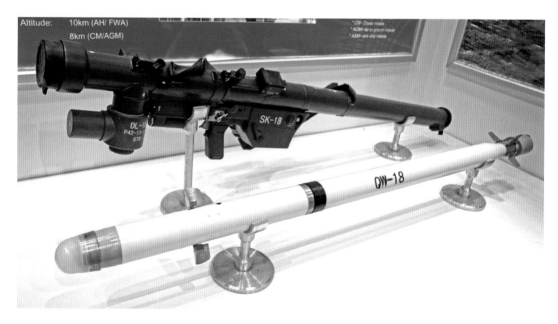

Defence Industry Exhibition (IDEX) show in Abu Dhabi. Designed to offer better resistance to IR jamming, it also has slightly greater slant range (6km vs 5km) and is fitted with a distinctive new BCU. In 2012, the QW-19 was unveiled at Airshow China in Zhuhai. The system adds an improved laser proximity fuze and may have an imaging-IR seeker as an option.

In parallel with the QW series developed under CASIC, the FN family (*Fei Nu*; Flying Crossbow) was developed by the 8th Research Institute of the China Aerospace Corporation (CASC) and manufactured by the Shanghai Academy of Spaceflight Technology (SAST). The first of the FN series debuted in 2002 as the FN-6. The seeker uses a multifaceted pyramidal dome very reminiscent of the French Mistral and probably incorporating a similar four-detector array. The FN-6 is offered with a two-blade IFF system similar to Stinger and Verba. Sudan claims to produce this missile under license as the Nayzak. The FN-16 debuted in 2008 at Airshow China in Zhuhai. This missile reverted to

The FN-6 is an outlier in Chinese MANPADS designs with a gripstock bearing little resemblance to Soviet/Russian types. In addition, the missile uses a multifaceted seeker cover more reminiscent of the French Mistral, perhaps indicative of a similar multidetector approach. (Author)

China's FN-16 Flying Crossbow was publicly unveiled in 2008 at Airshow China in Zhuhai. Compared to its predecessor, the FN-6, the FN-16 reverted to a more traditional MANPADS configuration with a hemispherical dome over the seeker. (Author)

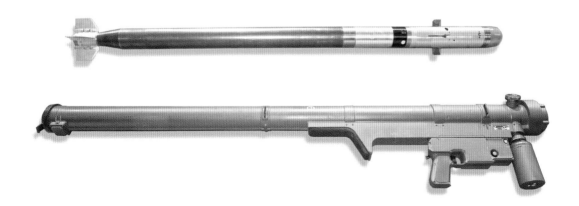

the more typical hemispherical domed seeker and offers better all-aspect engagement and better resistance against electronic countermeasures than the FN-6.

CHINESE MANPADS COMPARATIVE DATA					
	QW-1	QW-2	QW-18	FN-6	FN-16
Length	1.53m	1.59m	1.57m	1.49m	1.7m
Weight	16.5kg	18.4kg	18.3kg	17kg	18kg
Maximum altitude	4,000m	4,000m	4,000m	3,800m	4,000m
Slant range	5,000m	6,000m	5,500m	5,500m	6,000m
Debut	1994	1998	2007	2002	2008

ASIAN ALTERNATIVES

Japan is one of the handful of countries to manufacture an indigenous MANPADS, but Japanese MANPADS are some of the least known because Japan does not permit export of such weapons. The Type 91 Kin-SAM, originally called Keiko, began development in 1979. The seeker is relatively novel, based on a two-color IR/visible-light detector based around a high-resolution charge-coupled device. The sophistication of the seeker required a prolonged development by the Technical Research and Development Institute (TRDI), and the transition to engineering development at Toshiba did not begin until 1988. The first deployments with the Japanese Self-Defense Forces took place in 1994 and over 330 launchers were acquired. The Kin-SAM closely resembles the Stinger in layout, and is known as the "Hand Arrow" in Japanese service. An improved version called the Kin-SAM-2 Kai with improved imaging-IR guidance entered production in 2007.

South Korea's Agency for Defense Development (ADD) began development of a MANPADS in the late 1990s as the KP-SAM Shingung.

Toshiba's Type 91 Keiko Kin-SAM resembles the Stinger in basic outline but uses a different seeker design. This example is on display at Hamamatsu Air Base in Japan. Japanese MANPADS are not well known, because traditionally, they have not been exported. (Hunini/Wikimedia/CC BY-SA 4.0)

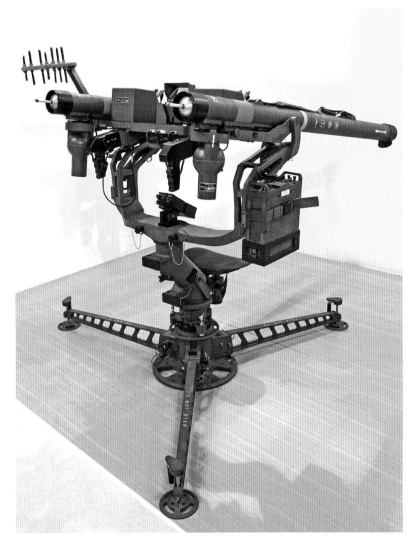

The South Korean electronics firm LIG Nex1 developed the KP-SAM Shingung based on technology transfers from the Russian Igla. As can be seen, its seeker cover features the aerospike pioneered by Soviet MANPADS. This is the Chiron export version on its standard pedestal launcher station. (Author)

It is manufactured by LIG Nex1 and is also called the Chiron in its export version. The missile is heavily based on Russian Igla technology obtained through a technology transfer program used to reduce Russia's outstanding commercial debt to South Korea. The KP-SAM became operational in 2005 and was publicly unveiled at the IDEX 2007 show in Abu Dhabi in hopes of winning export orders. The system is most commonly deployed on a tripod mount similar to the French Mistral rather than using a gripstock launcher.

North Korea's MANPADS are the world's most mysterious. North Korea manufactured what appears to be a license-built copy of the Soviet Strela-2M called the Hwaesung-Chong (Arquebus) although other accounts suggest it is a copy of the Chinese HN-5. There have been reports that the Soviet Union later sold North Korea the licensed-production rights for the Strela-3 and the Igla-1 system. In more recent years, North Korea has produced an upgraded Igla-1 with the system designation HT-16PGJ and the missile designation HG-16.

USE
MANPADS in combat

MANPADS OPERATION

The Redeye set the pattern for the worldwide development of MANPADS. Because most other types follow similar operating procedures, it is worth taking a brief look at how it functioned.

The Redeye was deployed in a two-man team consisting of a team leader and a gunner. The team leader, a sergeant, was responsible for communications with the HQ section, spotting targets, and directing the gunner. When alerted to an enemy aircraft, he removed the front protective cap from the launch tube, and folded the sight assembly into position. To begin the launch sequence, the gunner inserted a BCU into the receptacle at the base of the gripstock. The BCU contained the battery to operate the launcher as well as the coolant gas that was fed into the missile seeker to cool the detector. The gunner determined if the target aircraft was within range using the range ring at the front of the sight unit. Once the enemy aircraft was within range, the gunner pressed the actuator switch down with his right thumb to activate the BCU. The BCU had a 30-second lifespan, so it was not activated until the gunner was ready to fire. When the BCU was activated, it took about 3–5 seconds for the launcher's electrical and mechanical components to reach launch condition. Once ready, the launcher emitted a tone via a small acquisition indicator located on the sight unit near the gunner's ear. When the missile seeker sensed the target, the tone became progressively louder. As soon as the missile had locked on to the target, the gunner slid the uncaging switch forward using his left thumb. It was then up to the team leader to authorize the launch. Once given the command, the gunner would fire the missile using the trigger on the gripstock.

The Redeye did not have any built-in IFF system, so it was up to the team leader to determine the targets, based in large measure on information obtained through the air defense network. This usually came from a Target Alert Data Display Set (TADDS), a small electronic device linked to a forward-area alert radar (FAAR) by its built-in radio. The radar alerted the TADDS to the identity, location, and direction of enemy aircraft.

In US Army divisions, a Redeye air defense headquarters section was deployed at battalion level. This section managed the Redeye teams that were deployed at company level with four Redeye teams per battalion. The number of Redeye teams per division ranged from 49 to 62 depending on the organization. The equipment of each Redeye team varied. One of the more common deployments was to provide each team with an M151 ¼-ton 4×4 truck in order to provide the capacity to carry three missile/launchers as well as other team equipment.

COMBAT DEBUT: MID-EAST WARS

MANPADS saw their combat debut in 1969 during the War of Attrition (1967–70) fought by Egypt and Israel along the Suez Canal. The Soviet Union had supplied Egypt with high-altitude air defense missiles including the S-75 (SA-2 "Guideline") and S-125 (SA-3 "Goa") that saw extensive use in the opening phase of this conflict. The Israeli Air Force (IAF) countered these missiles by flying at low altitude, thereby exploiting the radar "dead-zone" of these missile systems. To cover this gap in coverage, in January 1969 a Soviet government delegation to Cairo promised that the Egyptian forces would be provided with the Strela-2 missile system. This took several months, due to the relative novelty of the Strela-2 as well as the need to train Egyptian crews to operate and maintain the weapons.

Egyptian crews were sent to the Soviet Union for training, and a special team from the KBM missile design bureau was sent to Egypt to monitor use of the Strela-2 missile. The first Egyptian Strela-2 teams were deployed in early August 1969 to protect S-75 missile batteries near the Suez Canal. The first successful use of the Strela-2 missile occurred on August 19, 1969, by the Egyptian 7th Strela Platoon covering an S-75M (SA-2) site 22km northwest of Suez City. This platoon claims to have shot down three IAF A-4H Skyhawks that day. Israeli accounts credit the loss of a single Skyhawk that day to antiaircraft gunfire. Subsequent Soviet accounts claimed that six IAF aircraft were downed during ten Strela-2 engagements in the late summer and early fall. This information was judged to be so important that Soviet premier Leonid Brezhnev was personally briefed in the Kremlin about the Strela-2's reported successes.

Israeli accounts attribute the first Strela-2 kill to an engagement with a Super Mystère on October 15, 1969. Regardless of the actual date of the first kill, the IAF changed tactics in the fall of 1969, warning its aircraft pilots to stay above 6,000ft danger zones to minimize their vulnerability to the new threat. There is some mystery as to whether it was Soviet specialists firing some of the Strela-2 missiles in these engagements due to a shortage of trained Egyptian gunners.

The discrepancies between Egyptian and Israeli accounts highlight a recurring problem with Strela-2 kill claims as well as other MANPADS kill claims such as those for the Stinger in Afghanistan. These missiles contain a self-destruct feature to prevent the missile from detonating on the ground should they miss their target. This is activated on a timer in the flight control system that detonates the warhead 14–17 seconds after missile launch at a range of 3–4km if the missile fails to hit the target. Overenthusiastic missile gunners, on seeing the missile self-detonate many kilometers away, attributed the explosion to a successful hit. Furthermore, not all Strela-2 hits were lethal. In some cases, the missile detonated near the targeted aircraft's tailpipe but the pilot was able to nurse his damaged aircraft back to base.

From the standpoint of the targeted aircraft, the cause of a loss is often unclear unless another aircraft is nearby and sees the missile impact or a stream of antiaircraft artillery (AAA) tracer rounds. Even in this case, it is often difficult to determine a missile strike. The rocket motors of most MANPADS burn out within a few seconds of launch, and contrary to movie depictions, the missile does not necessarily trail smoke in the seconds before impact.

The KBM missile design bureau claims that Strela-2 missiles destroyed 39 aircraft in the Egyptian theater in 1969–70. Other Russian military accounts indicate that Strela-2 missiles downed 15 IAF aircraft in 1969, while shooting down five aircraft and damaging two others in 1970. The lower kill claims in 1970 are attributed to the change in Israeli tactics, which kept the aircraft at higher altitudes during bombing missions. There were press accounts that at least one Strela-2 missile was recovered by the IAF in 1971 when it became lodged in the tailpipe of an Israeli jet after the impact fuze failed to detonate the warhead. This may account for the US development of IRCM during the time period. Reports in the United States estimated that about 100 Strela-2 missiles were fired against IAF aircraft during the War of Attrition.

By the time of the 1973 Yom Kippur War, Egyptian and Syrian forces had been supplied with as many as 2,000 Strela-2 and Strela-2M launchers. One US Department of Defense study estimated that about 5,000 missiles were fired during the course of the war; another study estimated 4,456 missile launches from 1,468 launchers accounting for the destruction of two or three IAF aircraft, 28 aircraft suffering tailpipe damage, and a few suffering minor engine damage. An article in the journal of the Russian air defense forces stated that Strela-2 missiles had accounted for six IAF aircraft during the conflict.

In general, the Strela-2 was viewed as having too small a warhead to ensure an aircraft kill; and by the time of the Yom Kippur War, Israel had deployed IRCM such as flares that degraded the performance of the Strela-2. However, the widespread proliferation of these weapons forced IAF strike aircraft to attack from higher altitudes than would have otherwise been the case in their absence, thereby degrading overall accuracy during bombing missions.

OPERATIONS IN VIETNAM 1972–75

The combat debut of the Strela-2 in the Republic of Vietnam took place during the Easter Offensive that began in March 1972. There had been reports of the use of a small antiaircraft missile by PAVN personnel as early as 1970, but these were generally discounted by US intelligence agencies because the PAVN regularly fired RPG-7 antitank rockets at low-flying aircraft and helicopters. The criterion for identifying the weapon in use as a guided missile was whether the missile actually maneuvered toward the target aircraft rather than following a ballistic path. From 1967 to 1970, there had been 380 recorded instances of US helicopters being fired on by RPG-2 and RPG-7 rockets with 128 helicopters shot down, and 30 that were hit but escaped.

There is some dispute about the precise date of the first use of the Strela-2 in Vietnam. A PAVN prisoner reported that a USAF OV-10A

Bronco observation aircraft shot down on April 1, 1972, had been the victim of a heat-seeking missile. Strela-2 sightings became increasingly common through April 1972. A US Army AH-1G Cobra attack helicopter was fired on by two missiles on April 3, 1972, but the launches were observed by a Forward Air Control (FAC) aircraft, and the helicopter pilot successfully outmaneuvered the missiles using a steep dive. On April 29, three F-4 Phantoms conducting low-level attacks at 500–1,000ft were fired on by ten missiles. The pilots managed to outmaneuver the missiles.

The first confirmed kill by a Strela-2 in Vietnam occurred on May 1, 1972, when a USAF O-2A Skymaster observation aircraft was shot down during the fighting around Quang Tri. A USAF A-1H Skyraider attack aircraft was shot down later the same day by two missiles, while a second A-1H was damaged in the same incident. The following day, two O-2As, two F-4 Phantoms, and an AC-130 gunship were engaged but outmaneuvered the missiles; but two A-1H Skyraiders and a UH-1H Huey helicopter were shot down on May 2. Numerous other encounters with the Strela-2 continued through May 1972, with the week of April 30– May 6 seeing about 30 missile launches, the highest weekly total during the war. Most of the incidents in 1972 occurred during the fighting around Quang Tri and An Loc, although there were a smaller number of incidents near Saigon and elsewhere. There were some encounters in North Vietnam near Hanoi, but the vast majority of the Strela-2 missiles were fired by PAVN units in South Vietnam.

The advent of Strela-2 missiles in Vietnam had a dramatic effect on air operations. The threat posed by these missiles led the USAF to restrict operations by propeller-driven FAC aircraft as well as helicopters and fixed-wing gunships because they were especially vulnerable to the Strela-2. Missions that had to be flown by these aircraft were generally conducted at altitudes of 9,000ft or higher to minimize the Strela-2 threat.

US intelligence had been aware of the Strela-2 threat due to the missile's use in the 1969–70 fighting along the Suez Canal. Efforts began immediately to equip larger aircraft with flare dispensers using the LUU-2/B IR flare. Helicopter crews used Very pistols with the Mk 50 IR flares. Starting on July 13, 1972, helicopters began to be fitted with "Toilet

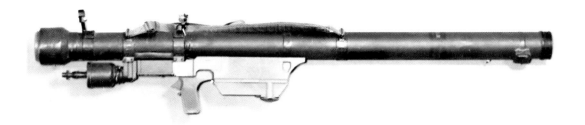

Bowl" exhaust diffusers that dispersed the exhaust gases into the propeller wash.

The most common tactical countermeasure to the Strela-2 was maneuvering, usually a maximum-effort bank into the missile's path. Maneuvering offered two means of defeating the Strela-2. It would often result in the Strela-2's seeker losing lock-on with the aircraft's exhaust. In addition, the Strela-2's propulsion burned for only 8.1 seconds after which the missile coasted, gradually losing energy. In the case of a maneuvering aircraft, attempts by the missile to maneuver could bleed off enough energy that it was unable to intercept. Most successful Strela-2 intercepts occurred while the missile propulsion was still active, on average around 5.7 seconds after launch.

A USAF study of Strela-2 engagements from April 1972 to January 28, 1973, recorded 350 engagements with 528 of the missiles. During these engagements, 45 aircraft were shot down and six were damaged; eight other aircraft were hit but not significantly damaged. Some form of countermeasure was used in 200 of the 350 engagements. Of the engagements with countermeasures, the most common was maneuvering alone (141); maneuvering with flares (44), and flares only (15). The Strela-2 missile was most effective against aircraft that failed to use any form of countermeasures: of 59 aircraft struck by Strela-2 missiles, 44 did not engage in countermeasures; 14 were against maneuvering aircraft and only one was against an aircraft using flares. Flares were not foolproof since they had to pass within the missile seeker's field-of-view in order to decoy the missile.

US forces withdrew from the conflict in Vietnam after the Paris Peace Accords were signed on January 27, 1973, but the Strela-2 continued to be used through the end of the Vietnam War in April 1975. The Army of the Republic of Vietnam was especially hard hit because the Strela-2, in conjunction with large numbers of antiaircraft guns, severely restricted helicopter support. An official USAF history later noted that:

Helicopters had less utility than in the 1972 offensive. With the build-up of SA-7s and AAA, helicopters could not operate in areas where North Vietnamese troops were deployed. Helicopter assaults were not feasible for restoring lost positions where SA-7s and concentrated AAA were deployed. As a consequence, much of the mobility that U.S. Army forces achieved by helicopters in the 1968 period was reduced in the 1972 offensive and almost completely withdrawn on the eve of the 1975 offensive ... The changing character

A Strela-2M in PAVN service during the concluding years of the Vietnam War. (USAF)

of the war – from a permissive air environment to a hostile one – neutralized the employment of helicopters except under select circumstances. (Lavalle 1977: 65)

The Republic of Vietnam Air Force (RVNAF) lost 84 aircraft, including 52 helicopters, in January–June 1974. Of these, 67 were attributed to AAA guns and 17 to SA-7 missiles. In total, the RVNAF lost 28 aircraft to SA-7 missiles from January 1973 to the end of 1974. Detailed RVNAF statistics of Strela-2 losses for the final four months of the war are lacking. The KBM missile design bureau claims that more than 205 aircraft were shot down in Vietnam during 1972–75. As mentioned earlier in the Egyptian case, kill claims by Strela-2 gunners tended to be exaggerated due to confusion between actual hits and explosions resulting from missile self-destruct.

The Strela-2 continued to be used in conflicts in South-East Asia. During the fighting along the Cambodian–Burma border in 1980–88, the Royal Thai Air Force was fired on no fewer than 58 times, with four aircraft hit by missiles. Two aircraft, an A-37B Dragonfly and an F-5E Tiger II, were shot down; another two A-37B Dragonflies survived.

AIR WARS IN THE AMERICAS

A variety of MANPADS were used during the 1982 Falklands War between Britain and Argentina. It was the only conflict that saw the widespread use of the Blowpipe missile – which, curiously enough, was in service with both combatants. More than 50 aircraft were shot down during the conflict, of which four have been credited to MANPADS. British units fired 95 Blowpipe missiles with roughly half suffering technical failures shortly after launch. Early reports claimed that the remaining missiles shot down nine aircraft and damaged two more. Later assessments trimmed this back to only one confirmed kill, an Argentinian Navy MB-339A shot down on May 28, 1982, during the fighting near Goose Green. The disappointing performance of the Blowpipe prompted Brigadier Julian Thompson, commander of 3 Commando Brigade, to describe its use as "trying to shoot pheasants with a drainpipe." The British SAS was equipped with the Stinger, and of about five fired, the Stinger was credited with shooting down an Argentinian Air Force Pucara light strike aircraft and an SA.330L Puma helicopter. Argentinian forces shot down a Royal Air Force Harrier GR.3 with a Blowpipe on May 21, 1982, during the fighting near Port Howard. Argentinian forces also had some Strela-2M missiles obtained from Peru, but do not appear to have had any successes with them.

The Nicaraguan Civil War of 1979–89 saw the first widespread use of MANPADS in Central America. In 1979, the Sandinista rebels (FSLN: Frente Sandinista de Liberación Nacional) overthrew the dictator Anastasio Somoza. The Nicaraguan Democratic Front, dubbed the Contras (counterrevolutionaries), began guerrilla operations against the Sandinista armed forces in the early 1980s. By the mid-1980s, the local civil war had become a proxy struggle, with the United States supporting the Contras and the Soviet Union and its allies supporting the Sandinista government. The Sandinistas received the Strela-2M starting in 1981 and by 1984 had acquired about 300 launchers. A 1987 CIA report indicated that Nicaragua later received Strela-3 and Igla-1 missiles as well.

The Strela-2 missiles were used against aircraft flying supplies to the Contra rebels, but they also were fired at A-37B Dragonfly strike aircraft of the Honduran Air Force during skirmishes along the contested border. The Contras also obtained the Strela-2M, reportedly from the CIA. On February 27, 1987, the Contras shot down a Sandinista Mi-24 "Hind" attack helicopter near San Pedro del Norte. In July 1987, the Nicaraguan government accused the CIA of supplying the Contras with Redeye missiles, which were credited with the downing of two Mi-24s earlier that year. A later US study concluded that about 38 SA-7 missiles had been fired by both sides during the civil war, with two transport aircraft and two helicopters shot down and seven other aircraft hit but surviving.

These weapons spilled over into neighboring conflicts. In late 1989, Strela-2M and Redeye missiles, presumably obtained in Nicaragua, turned up in the hands of leftist rebels in El Salvador.

During the 1995 Cenepa River War between Peru and Ecuador, at least five aircraft were shot down, including a Peruvian Air Force Mi-8 "Hip" helicopter struck by an Ecuadoran Blowpipe, a Peruvian Air Force

Mi-24 shot down by an Ecuadorian Igla missile, and an Ecuadorian Air Force A-37B Dragonfly shot down by a Peruvian MANPADS, probably an Igla. A rare example of the RBS 70 in combat took place on November 27, 1992, during an attempted *coup d'état* in Venezuela when a rebel OV-10A Bronco was shot down.

THE NAMIBIAN/ANGOLAN CONFLICT 1979–88

Conflicts stemming from various independence movements in Portuguese colonies led to the first extensive combat use of MANPADS in sub-Saharan Africa. Insurgents of PAIGC (Partido Africano para a Independência da Guiné e Cabo Verde; African Party for the Independence of Guinea and Cape Verde) obtained a supply of Strela-2 missiles from the Soviet Union to assist in fighting Portuguese forces in Guinea. The Portuguese Air Force lost three Fiat G.91s to the missiles, the first on March 25, 1973. The presence of the missiles forced the Portuguese Air Force pilots to avoid carrying out ground-attack operations below 8,000ft. The Strela-2 turned up shortly after in Mozambique with FPLM (Forças Populares de Libertação de Moçambique) insurgents. These missiles were not particularly effective due to the guerrillas' poor training and lack of fire discipline. When colonial rule ended in Spanish Sahara, Algeria supplied the Polisario Front with Strela-2M missiles that were used against Moroccan forces. They were credited with downing Royal Moroccan Air Force F-5E Tiger II and Mirage F.1 aircraft during the prolonged conflict.

After Portugal decided to end its colonial rule in 1974, Angola became embroiled in a civil war pitting three rival forces against one another: FAPLA (Forças Armadas Populares de Libertação de Angola; People's Armed Forces of Liberation of Angola), UNITA (União Nacional para a Independência Total de Angola; National Union for the Total Independence of Angola), and FNLA (Frente Nacional de Libertação de Angola; National Liberation Front of Angola). FAPLA was backed by the Soviet Union, which began arms shipments in 1975. Deliveries of the Strela-2 began in early 1975, followed by the Strela-2M in November 1975 along with air defense training teams.

In early 1976, the CIA negotiated with Israel for the transfer of about 50 Strela-2 launchers to the rival UNITA militias in exchange for 50 Redeyes. Cuban-piloted MiG-21 fighters were fired on by UNITA forces in March 1976 without success. In March 1977, UNITA managed to shoot down a few Angolan An-26 transport aircraft. UNITA forces shot down at least two more Angolan aircraft in 1979–80. The liberation wars fought in the various Portuguese colonies spilled over into southern Africa in the late 1970s, leading to the most intense battles in the region.

The South African Defence Force (SADF) had been engaged in a prolonged conflict in South West Africa with the People's Liberation Army of Namibia (PLAN), the armed wing of the South West Africa People's Organization (SWAPO) on its northwestern frontier since the late 1960s. The FAPLA victory in Angola led to armed support of the PLAN.

A UNITA soldier holds a Strela-3 captured during the fighting near Cuito Cuanavale in August 1987. This is the 9P59 transport/launch tube with the distinctive 9P51 BCU attached; the gripstock is missing. (Author)

In response, the South Africans intervened during Operation *Savannah* in October 1975. With Soviet support, the Cuban government escalated the confrontation in November 1975 with Operation *Carlota*, providing both military advisers and combat units to Angola. After South Africa seized much of South West Africa (Namibia), the conflict simmered for a few years with various cross-border incidents. FAPLA granted the Soviet Union basing rights for aircraft and warships in Angola, in exchange for increasingly sophisticated weapons including air defense missiles.

An Angolan Strela-2M shot down an SADF Impala strike aircraft on January 24, 1980, followed by a second on October 10, 1980. A PLAN unit shot down an SADF Alouette III helicopter on June 23, 1980. UNITA shot down an Angolan An-26 transport in November 1980 that was carrying Soviet advisers.

On August 23, 1981, the SADF launched Operation *Protea*, ostensibly to clear PLAN bases in Cunene province along Angola's southern border. The escalation of the conflict put the SADF in direct conflict with the Angolan FAPLA. On August 27, a South African Air Force (SAAF) Mirage III fighter was hit by a Strela-2M missile but managed to limp back to base with a damaged engine. There were several other encounters

57

between SAAF jets and Angolan missiles. By the end of the operation in September 1981, the SADF had captured 90–110 Strela-2M missiles. After a truce in 1981, the border fighting quietened but FAPLA stepped up its efforts to clear out the rival UNITA units in southern Angola. In December 1983, the SADF launched Operation *Askari* over the border, the first large-scale encounter with Cuban units. No SAAF aircraft were lost to the Strela-2M, but an Impala strike aircraft was downed by a 9K31 Strela-1 (SA-9 "Gaskin").

UNITA rebels continued to fire on Angolan and Cuban aircraft during the civil war inside Angola. A Cuban Mi-8 helicopter carrying Soviet advisers was shot down by a UNITA Strela-2M. Two An-12 transports and another Mi-8 were shot down later in 1984 by UNITA. On September 29, 1985, UNITA shot down an Angolan Air Force MiG-21MF fighter, one of the first fast jets knocked down by UNITA. The Soviet Union stepped up supplies of MANPADS to PLAN forces in 1985 including 150 Strela-2M launchers and 350 missiles, but PLAN soldiers did not prove to be very adept with their use.

In 1986, the United States began supplying small numbers of Redeye or Stinger missiles to the UNITA forces in Angola to deal with the threat of FAPLA attack helicopters. Russian accounts claim that Stingers were captured from UNITA as early as 1983, but there is no evidence that any Stingers had been delivered prior to 1986. The Soviet Union began supplying Angolan forces with the new Igla-1 starting in August 1987.

The SADF staged a major operation into Angola starting on August 14, 1987, near Cuito Cuanavale. This led to one of the largest battles of the conflict pitting the SADF against FAPLA and the Cuban expeditionary force. By this time, most Angolan infantry brigades had an organic MANPADS detachment with ten Strela-2M, Strela-3, or Igla-1 launchers. During the fighting, the Angolans managed to down a single SADF AM.3CM Bosbok spotter aircraft on September 3, 1987. The SADF captured a significant amount of Angolan equipment during this fighting, including some of the new Strela-3 missiles.

In 1988, an internationally brokered peace agreement led to Namibia's independence and ended the Angolan/South African fighting. During the course of nearly a decade of intermittent fighting, FAPLA, PLAN, UNITA, and Cuban forces had fired about 470 Strela-2 and Strela-3 missiles downing nine aircraft. In addition, seven aircraft were damaged. The fighting within Angola continued until 2002, with several more aircraft brought down by UNITA MANPADS.

MANPADS IN AFGHANISTAN

The Soviet Union's intervention in Afghanistan on December 24, 1979, led to a widespread insurrection by several Mujahideen insurgent groups. Owing to Afghanistan's mountainous terrain, air support became critical for Soviet military operations. The insurgents had acquired some Strela-2/Strela-2M missiles from Afghan National Army stockpiles but the Soviet Army quickly seized remaining inventories. The Mujahideen groups turned to Pakistan to obtain MANPADS. The Pakistani Inter-Services Intelligence (ISI) was the primary arms supplier to the Mujahideen. In 1980, the US government agreed to begin funding arms acquisition, with the ISI acting as the conduit to the Mujahideen. Details of the early supply of MANPADS to the Mujahideen remain sketchy, with both the ISI and the CIA apparently acquiring MANPADS through their own contacts. MANPADS began appearing in 1980, and the Soviet Air Force reported its first MANPADS loss on July 23, 1980, when an Mi-24 was shot down. MANPADS losses over the next few years were light: only one per year in 1980–82, three in 1983, and eight in 1984 according to Soviet accounts.

MANPADS SUPPLY TO AFGHAN INSURGENTS 1981–86						
	1981	1982	1983	1984	1985	1986
Strela-2	25	30	30	96	210	318
Blowpipe	0	0	0	0	0	28

Soviet reports indicate that one of the most significant supply sources was Egypt, with weapons supplied including the Strela-2, Strela-2M, and the locally built Ayn-al-Saqr. Other sources included Iran and China, with the Chinese HN-5 and HN-5A possibly coming from Pakistani stocks. According to Soviet accounts, the Mujahideen received about 3,000 MANPADS during the course of the war. The number operational at any one time was smaller: about 150 launchers in mid-1984, 341 in April 1987, and 691 at the end of 1988 according to Soviet accounts. In comparison, Soviet intelligence assessed the Mujahideen antiaircraft inventory in late 1988 to include 4,050 12.7mm HMGs and 770 tripod-mounted 14.5mm and 20mm guns.

The Strela-2 and its variants were not popular with the Afghan rebels due to their poor performance. Nevertheless, they were numerous enough to be a significant problem. In 1984 alone, the Soviet 40th Army captured 336 missiles and destroyed a further 330.

Some elements of the US government encouraged the supply of better MANPADS than Strela-2 variants, leading to a decision in 1982 to provide some of the older FIM-43C Redeye. The number of Redeyes sent to Afghanistan is not known but was probably small.

By the early 1980s, the Soviet Air Force had already developed a wide range of IRCM that had mixed results against the Strela-2 and its derivatives. The most effective were the PPI-26 and PPI-50 decoy flares ejected from the ASO-2V (helicopters) and ASO-28 (strike aircraft) dispensers. Early-generation MANPADS such as the Strela-2 were very susceptible to flares.

A Mujahid armed with a Strela-2M in Paktika province in 1987. Judging from the markings on the battery, this may be a license-built copy rather than a Soviet-built example. Many of the Strela-2M launchers in Afghanistan were from foreign sources including Egypt and China. (Private collection)

A Strela-2M team walking down a road in Logar province, Afghanistan, in 1984. The gunner in this photo was a former Afghan Air Force fighter pilot who shot down a Soviet Mi-8 "Hip" transport helicopter with a Strela-2M in spite of the helicopter employing decoy flares. (Private collection)

Another approach was the EVU (*ekranno-vykhlopnogo ustroystva*; exhaust-covering device) exhaust diffuser on Soviet helicopters. At first, the helicopter crews did not want to use the exhaust diffusers due to their drag and the resultant loss of performance; they became mandatory after 1983, however, due to increasing MANPADS losses. Finally, various types of "hot brick" IR jammers such as the SOEP-V1A Lipa (*Stanitsiya optiko-elektronnikh pomekh*; opto-electronic interference station) were introduced. These worked very well on some platforms such as helicopters but were less effective on fast-movers such as the Su-25 strike aircraft.

The United States reached an agreement with the United Kingdom over the purchase of Blowpipe missiles. The first Soviet report of a Blowpipe kill was on August 23, 1984, when a Mi-24 was shot down, followed by a second on November 7, 1985. However, most other accounts indicate that Blowpipe did not arrive in theater until September 1986 with an initial batch of 28 followed by about 225 more in July–August 1987. The Blowpipes were credited with at least two Mi-8 losses during October 1986–July 1987; and they were also credited with an attack on November 29, 1986, against an An-12 transport aircraft that departed Kabul and was hit at an altitude of 6,400m by a Blowpipe fired from a launcher located on a mountain top near the route. In spite of these accomplishments, the Blowpipe was not especially popular among the Mujahideen due to its weight and the difficulty using it. Soviet accounts suggest that a small number of the improved Javelin missiles were also sent to Afghanistan and this type is credited with downing a Mi-24P on March 24, 1987.

Prior to the introduction of the Stinger, the main causes of Soviet Mi-24 losses were HMGs. In 1985 the 12.7mm HMG was credited with 42 percent of the kills and the 14.5mm HMG with 25 percent. In the case of the Mi-8, the causes were 40 percent to 12.7mm HMGs, 27 percent to various small arms, 27 percent to autocannon such as the 14.5mm, and only 6 percent to MANPADS. Casualties in the strike aircraft regiments were somewhat different. In the Su-17 "Fitter" regiments, casualties through 1985 were attributed to 14.5mm HMGs (37.5 percent), 12.7mm HMGs (25 percent), MANPADS (25 percent), and small arms (12.5 percent).

Pakistan had already obtained the FIM-92A Stinger from the United States and Pakistan's leaders were encouraging the CIA to provide it to the Afghan insurgents. This had been resisted by several US government agencies including the CIA, the State Department, and the Joint Chiefs of Staff. The US Army was not keen on the idea, in part due to concerns that missiles were likely to fall into the hands of the Soviet Union and China, benefiting their own MANPADS research, as well as their development of IRCM. This resistance began to crumble in mid-1985 when the Soviet deputy director of the GRU (military intelligence) station in Athens defected and identified Greek employees of the local ITT subsidiary as having provided classified information on the Stinger, based on the European coproduction program. Besides the technological issues, there was also widespread concern in Washington DC that the missiles would invariably find their way outside Afghanistan and might be used against civilian airliners.

In the event, President Ronald Reagan decided to take more forceful steps against the Soviet Union in Afghanistan. In March 1985, he signed National Security Decision Directive 166 that authorized support of the Afghan insurgents "by all means available." This was not specifically directed at the supply of the Stinger, but included a broad gamut of military equipment and intelligence support. Opposition to the transfer of Stingers was gradually overcome by Congressional supporters. At a meeting of the Planning Coordination Group on February 25, 1986, formal opposition to the Stinger transfers was finally overcome, and transfers to UNITA in Angola were also approved. In March 1986, President Reagan signed a notification to Congress on the issue, ending the six-year barrier to the Stinger transfer signed during the previous Carter administration.

In June 1986, a team of Pakistani officers were sent to White Sands Missile Range, New Mexico, for training on the Stinger. These officers, led by Lieutenant Colonel Mahmood Ahmed of the ISI, would train Afghan insurgents in the use of the Stinger. The tempo of MANPADS use quickly escalated in 1984–86 as new MANPADS such as the Redeye, Blowpipe, and Stinger arrived in Afghanistan in growing numbers.

MANPADS USE IN AFGHANISTAN 1984–86: LAUNCHES VS SHOOT-DOWNS*			
	1984	1985	1986
Launches	62	147	847
Shoot-downs	8	10	23
Missiles per aircraft lost	7.75	14.7	36.8
*Based on Soviet accounts.			

A Mujahid Stinger gunner of the Mahaz-e Melli militia in the Sarkani district of Kunar province in 1987. This militia received 25 Stinger gripstocks and was credited with 33 hits on enemy aircraft. (AMRC)

The first recorded use of the Stinger in Afghanistan took place on September 25, 1986, near Jalalabad. Three missile teams of Gulbuddin Hekmetyar's Hizb-i-Islami militia set up an ambush for eight approaching Soviet helicopters. A Mi-24V of the 335th Separate Attack Helicopter Regiment piloted by A. Selivanin was attacked by three missiles, one missing the helicopter, one striking the rear compartment and killing the flight engineer, and the third starting a fire. Selivanin and the gunner parachuted to safety. A second Mi-24V piloted by Lieutenant E.A. Pogorelov was hit by two Stingers. Pogorelov ordered the other two crew members to parachute while he attempted to crash-land the badly damaged helicopter. Pogorelov was severely injured in the crash and died in hospital; he was posthumously awarded the Order of the Red Banner for his heroism. An accompanying Mi-8 was also shot down with only one crewman escaping. In October 1986, the aviation units supporting the 40th Army in Afghanistan lost ten aircraft to Stingers, three times the casualties of September. This included four of the well-protected Su-25 strike aircraft. The rugged Su-25 had a better chance of surviving a Stinger strike than most other Soviet aircraft. For example, on June 28, 1987, a Su-25 returned to base after a Stinger detonation had blown out its right engine.

Casualties in the Mi-24 squadrons continued to mount. On November 29, 1986, two two Mi-24s were downed by Stingers and two more on January 14, 1987. In the latter case, one of the downed helicopters was piloted by A. Selivanin who had previously been shot down in the first Mi-24 loss three months before. The growing Mi-24 losses led to an order to reduce the crew to two, the pilot and gunner, and not carry the usual flight engineer in the rear compartment.

As aircraft casualties mounted, the Soviet Air Force instructed the crews flying fixed-wing strike aircraft such as the Su-17 and Su-25 to operate at a minimum of 3,500–4,000m (11,500–13,000ft); helicopter crews adopted nap-of-the-earth flight patterns to stay under the Stinger's minimum effective altitude. Active steps were taken to reduce the vulnerability of strike aircraft, especially the Su-25. The Sukhoi chief

A Mujahid Blowpipe gunner of the Jabhah-yi Nejat-e-Melli militia in Kunar province in 1987. (AMRC)

designer, V.P. Babak, flew to Afghanistan on several occasions to inspect Su-25 aircraft that had survived Stinger strikes. The "hot brick" IRCM did not work well on the Su-25, forcing pilots to rely on a limited supply of flares. Sukhoi developed methods to harden the engine compartment to reduce its vulnerability to Stinger strikes. This led to the Su-25PBZh "increased combat survivability" variant that entered combat service in Afghanistan in August 1987.

Probably the most famous Su-25 pilot shot down by a MANPADS was Lieutenant Colonel Aleksandr Rutskoy, commander of the 378th Separate Attack Aircraft Regiment. Rutskoy's aircraft was hit during a mission near Zhawar on April 6, 1986, and he successfully ejected. There are conflicting accounts whether this was as a result of a Redeye or a Blowpipe hit. Regardless, Rutskoy survived a second shoot-down in August 1988 when his Su-25 was downed by a Pakistani F-16A Fighting Falcon near the border. He is best known for his role in the attempted coup against President Boris Yeltsin in October 1993.

Aside from aircraft IRCM and changes in aircraft and helicopter tactics, the Soviet 40th Army began an increased campaign of interdicting Mujahideen caravans along the Pakistan and Iran borders in the hopes of stopping the flow of MANPADS. These raids and ambushes were generally carried out by the special forces of the GRU intelligence service, better known by their acronym "Spetsnaz." The first Stingers were captured during a mission by the 186th Spetsnaz Detachment under Major E. Sergeyev near the town of Shah Joy.

STINGERS IN AFGHANISTAN 1986–89*					
	1986	1987	1988	1989	Total
Gripstocks delivered	36	227	45	0	308
Missiles delivered	154	1,111	323	0	1,588
Missiles launched	37	201	99	5	342
Missile hits	26	164	79	5	274
*Based on information provided by Lieutenant Colonel Mahmood Ahmed of the Pakistani ISI.					

By the time of the Soviet withdrawal from Afghanistan in February 1989, the Mujahideen claimed to have hit or shot down 274 aircraft and helicopters using Stinger missiles. Of these, 101 were helicopters, 92 were transport aircraft, and 81 were strike aircraft such as the Su-17 and Su-25. A US Army study based on data provided by the Mujahideen and ISI concluded that 269 aircraft were shot down in 340 engagements. On a related issue, a US study concluded that Strela-2 missiles in Afghanistan had downed 47 aircraft of which 42 were helicopters. Figures for Blowpipe and Redeye kills in Afghanistan are not readily available.

The success rate of the Stinger in Afghanistan became controversial for several reasons. To begin with, Stinger supporters in the bureaucratic battles over its supply to the Mujahideen saw the high kill figures as evidence of the decisive effect that the Stinger had in defeating the Soviet armed forces in the Soviet–Afghan War. The Soviet defeat in Afghanistan is still recalled with bitterness in Russia, and there has been little enthusiasm to declassify or publicize embarrassing data about Soviet losses in the war. As a result, there is reason for skepticism about both the Mujahideen claims and Soviet aircraft loss data.

Official Russian accounts state that total Soviet aircraft losses were 125 fixed-wing aircraft and 333 helicopters lost to all causes. There are some minor discrepancies in different source documents; for example, some sources state that 118 fixed-wing and 300 helicopters were lost. This difference may be due to inclusion or exclusion of aircraft not under 40th Army control such as the Border Guards. Figures for the Afghan Air Force are completely lacking.

There is no official breakdown of causes of the aircraft losses. Although it seems likely that the Soviet General Staff compiled such an analysis, it has not been publicly released. Russian aviation historians have attempted to reconstruct data on the losses. The cause of all losses is not complete, however, with about 6 percent still having an unknown cause. Accidents, pilot error, and other non-combat mishaps account for about 22 percent of the total. The remaining 290 combat losses fall into roughly three categories: ground fire, MANPADS, and aircraft destroyed at air bases during rocket attacks. The "ground fire" category includes small arms, 12.7mm, 14.5mm, and 20mm HMG/cannon as well as about

A Mujahid Stinger gunner of the Hezb-e-Islami militia in the Surkh-Rod district in 1989. This particular militia received the largest portion of Stingers distributed up to the end of the war, including 83 gripstocks. It was credited with 72 hits for 117 missiles fired. (AMRC)

nine helicopters shot down by RPG-7 rockets. This category is the most ambiguous because many aircraft losses are attributed simply to "enemy fire" without stating whether it was machine guns or MANPADS. As a result, the MANPADS category is probably undercounted. So for example, the Russian data upon which the chart below is based indicate that 25 Mi-24s were lost to MANPADS, but other Russian accounts indicate that 28 Mi-24s were lost to Stingers alone.

SOVIET AIRCRAFT COMBAT LOSSES IN AFGHANISTAN 1980–89											
Cause/type	1980	1981	1982	1983	1984	1985	1986	1987	1988	1989	Total
Ground fire, aircraft*	2	1	7	3	7	6	8	5	1	1	41
Ground fire, helicopter	22	14	15	11	23	32	20	11	5		153
MANPADS, aircraft				2	4	3	8	11	2	1	31
MANPADS, helicopter	1		1	1	5	5	11	18	6	1	49
Other, aircraft**					1				9	1	11
Other, helicopter							1	1	2		4
Aircraft losses, sub-total	2	1	7	5	12	9	16	16	12	3	83
Helicopter losses, sub-total	23	14	16	12	28	37	32	30	13	1	206
Total combat losses	25	15	23	17	40	46	48	46	25	4	289

* Fixed wing aircraft.
** "Other" was primarily aircraft destroyed on the ground by rocket/artillery attacks on air bases.

From this data, about 185 aircraft of all types were lost to ground fire (64 percent), and at least 80 to MANPADS (28 percent). MANPADS kills increased sharply in 1986–88 following the surge in MANPADS supplied, including both the Stinger and the Blowpipe. There is only fragmentary and contradictory Russian data on what percentage of Soviet aircraft were downed by the Stinger. Part of the problem is that after 1986, there was the tendency to attribute every MANPADS attack to the Stinger. For example, in the case of the Mi-24, some Russian accounts claimed that it was engaged by 563 Stingers with 89 hits and 18–28 kills, but this greatly exceeds the total number of Stingers launched against all types of aircraft during the fighting.

An Afghan Air Force Mi-8 "Hip" transport helicopter lost in Bamyan province in 1988. This helicopter is fitted with an EVU exhaust diffuser over the engine exhaust port, a feature introduced in 1985 to reduce the vulnerability of Soviet helicopters to the MANPADS threat. (AMRC)

There is a huge discrepancy between the figures given by Lieutenant Colonel Mahmood Ahmed, the ISI officer who headed the Stinger program in Afghanistan, when compared to published Russian accounts. The Mujahideen claimed to have shot down 274 Soviet and Afghan aircraft with Stingers, while Russian accounts suggest only about 80 MANPADS casualties for all types.

There are several reasons that partly explain the gap. Lieutenant Colonel Ahmed's account of the Stinger in Afghanistan sometimes refers to 274 hits or 274 aircraft shot down. There is a significant difference between "hits" and "shot down." Because in many cases Soviet aircraft were hit by multiple Stingers, it is possible that these figures actually reflect a smaller number of aircraft actually shot down versus 274 hit. As mentioned earlier, the Russian accounts probably undercount MANPADS losses due to the unknown causes for some losses, as well as loss reports that list "enemy fire" without specifying the type. Some aircraft counted as "kills" by the Mujahideen may have survived the missile impact and returned to base. The Soviet figures do not include Afghan Air Force losses nor civil transport aircraft. The Afghan Air Force was relatively small compared to the Soviet Air Force presence, so its losses would narrow the gap, but not completely close it. At least four airliners and transport aircraft of Bakhtar Afghan Airlines and Bakhtar Alwatana Airlines are known to have been shot down by MANPADS.

The total number of aircraft downed by the Stinger in Afghanistan was not the only indicator of its combat effectiveness. The sudden rise in strike aircraft and helicopter losses to MANPADS in late 1986–early 1987 forced the Soviet Air Force to severely curtail close-support missions. The use of Mi-8/-17 transport helicopters was significantly limited due to their vulnerability, dramatically affecting the ability of the Soviet 40th Army to conduct heliborne operations. The better-protected Mi-24 attack helicopter was limited to standoff ranges over 3,500m, which significantly degraded its ability to conduct fire-support missions. Likewise, missions by strike aircraft such as the Su-17 and Su-25 were restricted to altitudes over 3,500m, compromising their effectiveness. Compared to the USAF and US Navy in Vietnam, the Soviet Air Force in Afghanistan did not make extensive use of precision guided weapons such as laser-guided bombs during this conflict. This limited the ability of their strike aircraft to conduct standoff attacks and forced them to rely on less accurate alternatives such as unguided rockets. Soviet ground troops began referring to Soviet pilots as "Cosmonauts" due to their attacks from great altitudes. As mentioned in the introduction of this book, MANPADS can cause virtual attrition against an opposing air force by degrading their ability to conduct their missions. The Stinger in Afghanistan provides a classic case of virtual attrition.

Proponents of the delivery of the Stinger to Afghanistan have claimed that it was the Stinger that drove the Soviet forces out of Afghanistan. This is certainly an exaggeration. However, the Stinger did substantially degrade the Soviet employment of helicopters and strike aircraft, taking away one of the principal Soviet tactical advantages against the Mujahideen forces. Seldom does a single weapon so dramatically affect a war's outcome as was the case with the Stinger in Afghanistan.

Following the Soviet withdrawal from Afghanistan, fighting continued in the country between the new Taliban government and various regional warlords. From 1989 until the US intervention in September 2001, a total of about 60 fixed-wing aircraft and over two-dozen helicopters were shot down by various factions, with at least 15 attributed to MANPADS including remaining Stinger missiles.

THE PROLIFERATION PROBLEM

The threat of the use of MANPADS by non-governmental armed groups started in the early 1970s. In 1973, Italian police thwarted an attempt by the Palestinian Black September organization to shoot down an Israeli Boeing 707. There is some dispute over the nature of some of these attacks because civilian airlines are often used by the military for transporting personnel and supplies. For example, an Air Vietnam Douglas C-54D was shot down by a PAVN Strela-2 on March 12, 1975, but this incident falls into a gray area between military and civilian conflicts.

The first shoot-down of an airliner by a non-governmental armed group occurred on January 29, 1978, when a Chadian Air Force Douglas DC-4-1009 was hit by a Strela-2 fired by the National Liberation Front of Chad. Most subsequent MANPADS attacks occurred in war zones. For example, the Zimbabwe People's Revolutionary Army (ZIPRA) shot down two Air Rhodesia Vickers Viscount 782Ds in 1978–79 using Strela-2 missiles. UNITA made at least 13 attempts to shoot down Angolan and various other airliners from 1983 to 2001, hitting seven and missing six. The single costliest attack was the UNITA shoot-down of a

TAAG Angola Airlines Boeing 737-2M2 on November 8, 1983, killing all 130 passengers and crew.

There were numerous Mujahideen attempts during the Soviet–Afghan War to shoot down aircraft of Bakhtar Afghan Airlines. One of the reasons the United States was reluctant to ship Stingers to the Mujahideen was the fear that some would be used against civilian airliners. In the wake of the Soviet–Afghan War, the CIA went to considerable lengths to retrieve unused Stinger missiles, in many cases paying a bounty for their return.

There were several incidents in Sudan and Chad in the 1980s and 1990s during the civil wars there. In 1993 during the fighting between Abkhazia and Georgia, Abkhazian forces made at least five attempts against Georgian airlines, shooting down a Transair Tupolev Tu-134A-3 and an Orbi Georgian Airlines Tupolev Tu-154B on September 20 and 22, respectively.

From a geographic perspective, Africa was the hot spot for attacks on civilian airliners, with 64 percent occurring there. The most dangerous locales in terms of the number of incidents were Angola, Sudan, Afghanistan, Rhodesia, and Georgia. Nearly all of these attacks were linked to civil wars, insurrections, and other conflicts; only about 10 percent of attacks occurred outside conflict zones.

MANPADS attacks against civil air transport peaked in 1986–93. In total, there were about 65 incidents from 1973 to 2017 with approximately 1,000 passengers and crew killed. In the majority of cases, the type of missile used is not known. In the cases where the missile was identified, most were the Strela-2 or Strela-2M. Turboprop aircraft suffered the most fatalities; jet airliners tend to be larger and have a better capacity to return to the airport if a single engine is disabled.

Perhaps the most consequential misuse of MANPADS occurred on the evening of April 6, 1994, when the Dassault Falcon 50 business jet carrying Rwandan president Juvénal Habyarimana and Burundian president Cyprien Ntaryamira was shot down by two Igla-1 missiles as it prepared to land in Kigali, Rwanda. The assassinations precipitated the Rwandan genocide, leading to the deaths of hundreds of thousands of Rwandans, and a prolonged civil war. The Igla-1 missiles used in the incident appear to have come from Ugandan arsenals but the identity and motives of the perpetrators remains a very controversial mystery to this day.

Growing alarm over the threat posed by the leakage of MANPADS to non-governmental armed groups was addressed the Wassenaar Arrangement in 1996, approved by 42 countries including Russia and the United States. Wassenaar members agreed on non-binding criteria to guide exports of MANPADS. Of the major MANPADS manufacturers, China is not a signatory of the agreement. The United States started a program in 2003 to eliminate world-wide stocks of MANPADS. The US State Department oversaw the destruction of more than 40,000 missiles, mainly from former Warsaw Pact armies and elsewhere in East–Central Europe.

One method to skirt around the MANPADS export limitations is to mount the missile system on some form of pedestal launcher, making it too cumbersome for actual man-portability. For example, Russia sold Igla missiles to Libya and Syria but with the Strelets multiple-launch system and not with lightweight gripstocks.

Following the withdrawal of Soviet forces from Afghanistan in 1989, the CIA conducted a program to retrieve unused Stinger missiles. This is the launch tube from a Stinger missile expended in Afghanistan and currently on display at the CIA Museum at Langley, Virginia. (US CIA)

One reason for the decline in the numbers of MANPADS attacks may have been the gradual disappearance of the old first-generation systems. Newer generations of MANPADS with cooled seekers require a BCU. These become age-expired after several years, and so the missile launcher is no longer functional without one. The first-generation MANPADS with uncooled seekers require only a battery. While the system-specific batteries age-expire, there are improvised means to operate these launchers using other electrical sources, whereas BCUs are much more difficult to improvise if lacking very sophisticated technical support. The missiles themselves have age issues, such as the degradation of the solid-fuel rocket motors. The typical Soviet-era MANPADS had a warranty life of ten years. There have been cases of missiles over 20 years old being successfully fired, but most missiles increasingly will fail to launch if they are beyond their warranty dates.

IRAQ AT WAR
For a variety of reasons, MANPADS do not appear to have played a major role in the bloody Iran–Iraq War of 1980–88. Iraq had some Strela-2M missiles at the beginning of the conflict, but they do not appear to have had much of a role against the Islamic Republic of Iranian Air Force.

Iran apparently had a modest program for licensed production of the Strela-2M, but this was heavily dependent on the supply of key components from the Soviet Union. The Strela-2M was used by Iranian forces to defend high-value targets such as oil rigs and refineries. There have been reports that an Iranian Strela-2M managed to shoot down a P-15 (SS-N-2 Styx) antiship missile that was fired against an oil platform. Iran obtained additional MANPADS during the course of the war, with Libya and Syria being major sources. There have been reports that missiles manufactured in the Soviet Union were more reliable than those "built in Asia," presumably Chinese HN-5 missiles. The Strela-2M was widely used in the final stages of the war. Although they did not shoot down large numbers of Iraqi aircraft, they did succeed in dissuading Iraqi strike aircraft and helicopters from making close firing runs against Iranian positions.

The French Army deployed three Mistral-equipped sections during Operation *Daguet,* the French portion of the coalition war against Iraq in 1991. These sections were primarily assigned to defend artillery regiments. (US DoD)

Iran bought ten RBS 70 launchers and 400 missiles in 1985 under dubious circumstances in spite of the arms embargo against Iran and Sweden's tight export laws. They saw their combat debut in January–February 1987. Of the 42–45 Iraqi aircraft kills credited to Iranian MANPADS in the area east of Basra, the majority were credited to the RBS 70. Iran apparently obtained more missiles from international shell companies, leading to a scandal in Sweden. Overall, MANPADS do not appear to have had a major impact on the conduct of the Iran–Iraq War, and were certainly not as influential as in Vietnam or Afghanistan.

By the time of the 1990–91 Gulf War between Iraq and the international coalition, Iraq had obtained additional Strela-3 and Igla MANPADS. The coalition forces lost a total of 38 fixed-wing aircraft to Iraqi actions of which nine were credited to antiaircraft gunfire, 13 to MANPADS, and ten to radar-guided missiles. The majority of the losses to MANPADS involved close-support aircraft including six USAF A-10A/OA-10A Thunderbolt II aircraft, three US Marine Corps AV-8B Harrier IIs, and two US Marine Corps OV-10D Bronco observation aircraft. The only other coalition aircraft lost to MANPADS was a Royal Air Force Tornado GR.1. Some of these MANPADS losses may have been to larger, vehicle-launched missiles such as the Strela-1 and 9K35 Strela-10 (SA-13 "Gopher").

The second Iraq War starting in 2003 saw far fewer aircraft losses to MANPADS, largely because much of the Iraqi inventory was age-expired. One Iraqi officer recalled that in one day in April 2003, he fired a dozen Strela-3/Igla-1 MANPADS at US aircraft; all failed to launch. These had all been obtained in the same 1983 contract and so were well beyond their warrantied life. The Iraqi MANPADS arsenal was refreshed by supplies from Iran. From 2003 to 2009, coalition forces lost a total of 48 helicopters and three fixed-wing aircraft to hostile fire. Of these, eight helicopters were shot down by MANPADS though some other helicopter losses attributed to ground fire may have been due to MANPADS.

THE BREAK-UP OF YUGOSLAVIA

During the 1990s, Yugoslavia underwent a prolonged civil war that led to the country's gradual breakup. War broke out in Slovenia in June 1991 quickly leading to its independence. Croatia rapidly followed suit and the war between the Serbian Army and Croatia formally ended on January 3, 1992. In April 1992, a major war broke out in Bosnia-Herzegovina, pitting the self-proclaimed Serbian Republic of Bosnia, backed by Serbia, against local forces. On September 3, 1992, an Italian Air Force G.222 transport aircraft was shot down, presumably by a Strela-2M, while approaching Sarajevo airfield on a United Nations relief mission. Serbian forces shot down a Croatian Air Force MiG-21 with a Strela-2M in September 1993. On April 16, 1994, a Royal Navy Sea Harrier FRS.1 operating from HMS *Ark Royal* was shot down by a MANPADS of the Bosnian-Serb Army (VRS: Vojska Republike Srpske) while attempting to bomb Bosnian-Serb tanks near Gorazde. On 17 December 1994, a French Navy Etendard IVP was hit by a Strela-2M but managed to return to its carrier.

NATO became more deeply involved in the conflict in August 1995 following Serbian shelling of Sarajevo. A combined NATO force of about 400 aircraft were used in the three-week campaign, dubbed Operation *Deliberate Force*, that set about to undermine the military capability of the Bosnian-Serb Army. The Serbian forces had an ample supply of MANPADS because the former Yugoslavia had locally manufactured the Strela-2M, including the improved Strela-2M/A. Serbia also appears to have obtained more modern types such as the Igla from former Soviet republics such as Kazakhstan. The large number of MANPADS in Serbian hands led to a NATO tactic of restricting air missions to medium altitudes specifically to avoid the MANPADS threat. Nevertheless, on August 30, 1995, a French Air Force Mirage 2000N was shot down by a Bosnian-Serb MANPADS at an altitude of about 3,000ft and the crew captured. This was the only aircraft lost in the campaign due to hostile fire.

In 1998, significant fighting broke out in the Kosovo region, between ethnic Albanian and Serb military units. In March 1999, NATO initiated a major air campaign called Operation *Allied Force* that aimed to eject Serbian forces from Kosovo. As in the case of the previous air campaign, the presence of large numbers of Serbian MANPADS led the NATO air forces to conduct their campaign from altitudes over 15,000ft. NATO eventually had to relax these rules to permit aircraft to verify their targets. During the air campaign, Serbian forces fired about 700 missiles against NATO aircraft, but these were primarily large radar-guided missiles such as the S-125 (SA-3 "Goa"). An Igla-1 was credited with the loss of a USAF F-16C Fighting Falcon over Kosovo on May 2, 1999. MANPADS played an important role in shaping NATO tactics even though they were not fired in large numbers.

WARS IN RUSSIA'S "NEAR ABROAD"

Following the collapse of the Soviet Union, Russia was involved in a string of conflicts in "The Near Abroad"– the former Soviet republics and Russia's border regions. The initial war in Chechnya from December 1994 through August 1996 was the costliest of these conflicts. The Chechen forces had access to old Soviet arsenals including Strela-2M and Igla MANPADS, as well as personnel previously trained in the Soviet Army. Russian aircraft losses were about 38 of which 15 were caused by MANPADS and ten were probably MANPADS, with the remainder to various other types of antiaircraft weapons. This suggests that about 66 percent of the losses were due to MANPADS. During the Second Chechen War (1999–2007), Russian forces lost a further 45 helicopters and eight fixed-wing aircraft, the majority of which were probably caused by MANPADS.

One of the most prolonged conflicts in the former Soviet republics was between Armenia and Azerbaijan over the contested Nagorno-Karabakh region. During the first wave of fighting in 1988–94, about 21 aircraft were shot down of which at least four were caused by MANPADS as well as several others lost due to unknown hostile action but possibly MANPADS. There was another outbreak of serious fighting in 2021.

During the 2008 war between Russia and Georgia over the disputed Ossetia region, Soviet-era MANPADS were used by all sides. In addition, Georgia acquired an Igla derivative, the Polish Grom, including 30 gripstocks and 100 missiles. The first Russian aircraft loss was an "own goal" on August 8, 2008, when the Su-25 piloted by Lieutenant Colonel Oleg Terebunsky of the 368th Attack Aviation Regiment was shot down near the Zarsk pass in Ossetia by a MANPADS missile fired by allied South Ossetian militiamen. The following day, another Su-25 piloted by Colonel Sergey Kobylash, commander of the 368th Attack Aviation Regiment, was hit by a Georgian MANPADS that disabled his left engine. Moments later, his aircraft was struck by a second MANPADS, apparently fired by the South Ossetian militia. The loss of two Russian aircraft to friendly fire was a major incentive in the redesign of the IFF system for the new Verba MANPADS. Polish sources claim that the Georgian Grom missiles were responsible for downing nine Russian aircraft and helicopters of the roughly 20 shot down during the shot conflict. Some Grom systems captured by Russian forces in Georgia turned up in the conflict in eastern Ukraine in use with separatist militias.

MANPADS have been used extensively in the conflict in eastern Ukraine. In one of the costliest incidents, a Ukrainian Air Force Il-76MD "Candid" transport aircraft was shot down by a MANPADS fired by Ukrainian separatists near Lugansk on June 14, 2014, killing the 59 men on board. In the first two years of the conflict in 2014–15, nine combat aircraft, three transports, and ten helicopters were shot down, primarily by Igla missiles.

MANPADS again saw extensive use in the 2022 war in Ukraine. This conflict saw the combat debut of a number of MANPADS supplied by NATO countries to Ukraine such as the Polish Piorun and British Starstreak, as well as use of other types not previously seen in this conflict including the Stinger and Mistral. At the time of writing, not enough information had appeared to draw any definitive conclusions. The extensive use of MANPADS by Ukrainian forces seems to have caused a significant number of Russian aircraft and helicopter losses, however, discouraging close air support.

THE SYRIAN CIVIL WAR

The Syrian Civil War (2012–20) saw the Syrian Air Force conduct numerous strike missions against rebel positions, but very few of its aircraft were lost to MANPADS. The first reported loss of a government aircraft to an insurgent MANPADS was in November 2012 near Aleppo. Given the scale and duration of the fighting, the number of aircraft lost to MANPADS was small. Various accounts of the Syrian Civil War have suggested that about a dozen Syrian military aircraft were shot down by MANPADS during the eight years of conflict.

Syria had obtained large quantities of Strela-2M missiles from the Soviet Union, but by the time of the civil war, most were decades beyond their warranty. Rebel groups obtained some of these missiles from captured government stockpiles and attempted to rejuvenate them using improvised batteries. These efforts had limited results because the missiles were so old that other components of the missiles and launchers, such as the rocket engines and gas-generators, were no longer functional. For example, the Ahfad al-Rasul Brigade captured about 50 Strela-2M missiles at the 46th Regiment base near Aleppo in late 2012, but none worked. This militia also captured some Igla-1 missiles from Syrian stockpiles. Of five Igla-1 missiles they attempted to launch, only one functioned, but it was credited with downing a Syrian aircraft. The Ahfad al-Rasul Brigade claimed at least one combat aircraft and a helicopter downed near the Abu ad-Duhur air base.

Aside from Syrian government stockpiles, the insurgents began to receive small quantities of MANPADS from sympathetic governments in the region. The Chinese FN-6 began showing up in Syria in February 2013, variously attributed to Qatar or Sudan. Some more recent Russian types also appeared in Syria, including the Igla-1 and Igla-1M. Even a few North Korean HJT-18 missiles showed up.

The Russian Air Force intervened in the conflict in 2015 and flew more than 39,000 sorties though 2018, far more than the battered Syrian Air Force. The Russians were well aware that the Syrian rebel forces had MANPADS, and tailored their combat missions to minimize the threat. Of 18–19 Russian combat aircraft losses in Syria through 2018, there was only one confirmed loss of a Russian combat aircraft to MANPADS, a Su-25SM shot down near Idlib on February 3, 2018, plus another damaged on October 1, 2015. The Turkish Army lost at least one helicopter to MANPADS during the fighting along the border. International pressure to limit MANPADS proliferation was a major factor in suppressing the use of these missiles in Syria. The United States refused to supply Stingers to the insurgents, and applied pressure on allied governments to enforce the embargo.

CONCLUSION

Even though newer MANPADS with cooled seekers are not as vulnerable to IR flares, these countermeasures remain a common feature in war zones due to the remaining arsenals of older MANPADS such as the Strela-2M and its derivatives. This is a demonstration firing from a Kamov Ka-52 "Hokum-B" attack helicopter over Kubinka air base in Russia in 2016. (Author)

MANPADS had their most dramatic impact in their first two decades on the battlefield, starting with the 1970–73 Middle East wars, the Vietnam War in 1972–75, the wars of decolonialization in Africa, and the Soviet–Afghan War in the 1980s. During these conflicts, air forces were only just beginning to adopt IRCM and tactics to circumvent the MANPADS threat. Both the United States and the Soviet Union recognized the vulnerability of early MANPADS to IRCM and began fielding such countermeasures shortly after the advent of MANPADS on the battlefield. By the 1990s, the technical limitations of MANPADS were better understood by air forces, and they could be countered both by technology and tactics.

Even if new forms of IRCM have blunted the MANPADS threat, the ubiquitous nature of these weapons has left a lasting imprint on the modern battlefield. Air forces can no longer attack ground forces with impunity. Some traditional forms of air attack, such as strafing ground targets with machine guns, are prohibitively risky in the presence of MANPADS due to the inherent short range of such guns. No IRCM are foolproof, however, and so air forces over the modern battlefield have tended to skirt the MANPADS threat by staying at higher altitudes to minimize the threat. The necessity to strike ground targets from higher altitudes corrodes the combat potential of air strikes. The power of MANPADS to dissuade close air support was most clearly the case during NATO's intervention in the Balkans in the 1990s, when NATO refrained from the use of attack helicopters due to the MANPADS threat, and during the 2022 war in Ukraine where Ukrainian MANPADS dissuaded the Russian use of attack helicopters for close support.

A member of the Free Syrian Army prepares to fire a Chinese FN-6 MANPADS system against Assad regime forces' aircraft conducting airstrikes in Aleppo, Syria, on October 13, 2014. (Ahmed Hasan Ubeyd/Anadolu Agency/Getty Images)

The lingering effect of the MANPADS threat is to oblige attack helicopters and strike aircraft to routinely carry hundreds of pounds of flare dispensers, IR jammers, and other countermeasures. Every pound of countermeasures carried is a pound less of payload that can be carried. Although fewer aircraft are shot down by MANPADS, the degradation of aircraft payloads due to the increasing cost and weight of missile countermeasures imposes virtual attrition on modern air forces.

The interplay between MANPADS and aircraft technology is not yet over. The advent of true imaging seekers on MANPADS will make many forms of IRCM obsolete, especially flares and IR jammers. These advanced seekers cannot be easily bluffed. While some new countermeasures such as directed-IRCM systems can defeat even the most sophisticated seeker, these systems are extremely costly, very heavy, and have not proven to be reliable enough even after more than 20 years of development. The Wizard War between MANPADS and countermeasures is likely to continue indefinitely.

Besides countermeasures and maneuvering tactics, air forces have countered the MANPADS threat with the development of standoff weapons such as precision guided missiles and bombs, which can be successfully employed with great accuracy from beyond the range of MANPADS. Yet these weapons are considerably more costly than attacking ground targets with machine-gun strafing and "dumb" bombs. There is no easy response to standoff tactics in the future development of MANPADS because these small missiles are inherently limited in size and weight. Greater range requires more powerful rocket engines, which in turn make the missiles too heavy to be man-portable.

MANPADS are also evolving to deal with new adversaries such as the new threat of small attack drones. Contemporary MANPADS such as Verba and the latest versions of Stinger have improved proximity fuzes that allow them to detect and defeat even small targets.

MANPADS will continue to be ubiquitous on the modern battlefield. In the case of insurgent groups and poorly equipped armies, MANPADS are one of the best counterweights to even the best-equipped air forces. Even aircraft and helicopters with sophisticated countermeasures are not always effective, and so MANPADS remain a useful shield to protect ground troops from the threat of close air attack.

GLOSSARY

AMRC	Afghan Media Resource Center
BCU	Battery/coolant unit
CIA	Central Intelligence Agency (United States)
IFF	Identification Friend or Foe
GAU	Glavnoye artilleriyskoye upravleniye (Main Artillery Directorate)
GOI	Gosudarstvenniy opticheskiy institut (State Optical Institute); in Leningrad/St. Petersburg
GRAU	Glavnoye raketno-artilleriyskoye upravleniye (Main Missile and Artillery Directorate)
IR	Infrared
IRCM	Infrared countermeasures
KBM	Konstruktoskoe byuro machinostroeniya (Industrial Design Bureau), formerly SKB GA; in Kolomna
KP-SAM	Korean Portable Surface-to-Air Missile
LOMO	Leningradskoe optiko-mehkanicheskoe obyednenie (Leningrad Optical Industrial Association)
MANPADS	Man-portable air defense system
NIZAP	Nauchno-ispaytatelniy zenitno-artilleriyskiy polygon (Anti-aircraft Artillery Research-Testing Proving Ground); in Donguz
OKB	Opytnoe-konstruktorskoe biuro (Design-development office)
ooSpN	otdelniy otryad Spetsialnogo Naznacheniya (Separate Special Operations Detachment)
PAVN	People's Army of [North] Vietnam
RMP	Reprogrammable microprocessor
RVNAF	Republic of [South] Vietnam Air Force
SKB GA	Spetsialnoe konstruktoskoe byuro gladkostvolnoy artillerii (Special design bureau of smoothbore artillery); later KBM

SELECT BIBLIOGRAPHY

The bibliography here primarily covers MANPADS. Besides published works, the author relied on his large collection of manufacturers' brochures collected at international airshows and arms exhibitions. The author also used many historical accounts of air campaigns and technical manuals that are too numerous to list.

Ahmed, Mahmood (2012). *Stinger Saga: How the Air Battle was Fought and Won in Afghanistan.* Bloomington, IN: Xlibris.

Anonymous (1973). *Evaluation of the SA-7 Surface-to-Air Missile Firings in SEA.* San Antonio, TX: USAF Special Communications Center.

Anonymous (1993). *China Today: Defense Science and Technology, Vol. 2 Part 3: Conventional Weapons and Equipment.* Beijing: National Defense Industry Press.

Artamonov, Stanislav, et al. (2006). "Shaytan-arba pod ognem: Poteri i povrezhdeniya Mi-24 v Afganistane," *Aviatsiya i Vremya* 5: 40–45.

Ashkenazi, Michael, *et al.* (2012). *MANPADS: A Threat to Civilian Aviation?* Bonn: Bonn International Center for Conversion.

Bogdanov, Andrey (2018). "Primenenie PZRK v Angole (boevye episody)," *Tekhnika i Vooruzhenie* 2: 26–32.

Cagle, Mary (1975). *History of the Redeye Weapon System*. Redstone Arsenal, AL: US Army Missile Command.

Duske, Heiner, *et al.* (1998). *Experimental Flak-Weapons of the Wehrmacht Part 2*. Neumünster: Nuts & Bolts.

Ginor, Isabella & Remez, Gideon (2017). *The Soviet–Israeli War 1967–1973: The USSR's Military Intervention in the Egyptian–Israeli Conflict*. Oxford: Oxford University Press.

Kuperman, Alan J. (1999). "The Stinger missile and U.S. intervention in Afghanistan," *Political Science Quarterly*, Summer 1999: 219–63.

Lavalle, Arthur J.C., ed. (1977). *The Vietnamese Air Force, 1951–1975: An Analysis of its Role in Combat and Fourteen Hours at Koh Tang*. Washington, DC: US Government Printing Office.

Lavrov, Anton (2018). *The Russian Air Campaign in Syria: A Preliminary Analysis*. Washington, DC: Center for Naval Analyses.

Lindenmuth, James (1972). *US Army Helicopter Operations in Vietnam and the Effects of the Rocket Propelled Grenade*. Aberdeen, MD: US Army Materiel Systems Agency.

McManaway, Major William (1990). "Stinger in Afghanistan," *Air Defense Artillery*, January–February 1990: 3–8.

Markovskiy, Viktor (1994–98). "Zharkoe nebo Afghanistana," 12-part series, *AeroKhobbi* 3–5.

Markovskiy, Viktor (2007). "Aviatsiya Spetznaza (Chast 12)," *Tekhnika i vooruzhenie* 7: 43–47.

Mercillon, Patrick (2018). *Missiles européens au combat*. Nanterre: ETAI.

Pawloski, Richard (1988). *Fighter Weapons Symposium, Book III: Surface-to-Air Threats*. Fort Worth, TX: General Dynamics.

Petukhov, S.I. & Shestov, I.V. (1997). *Istoriya sozdaniya i razvitiya vooruzheniya i voennoy tekhniki PVO Sykhoputnykh Voisk Rossii*. Moscow: Izd. VPK.

Schroeder, Matt (2014). "Fire and Forget: The Proliferation of Man-Portable Air Defense Systems in Syria," *Small Arms Survey Issue Brief 9*.

Shirokorad, A.B. (2003). *Entsiklopediya otechestvennogo raketnogo oruzhiya 1817–2002*. Minsk: Kharvest.

Singh, Mandeep (2020). *Air Defence Artillery in Combat 1972 to the Present: The Age of Surface-to-Air Missiles*. Barnsley: Pen & Sword.

Sukolesskiy, A.V. (2009). *SPETsNAZ GRU v Afghanistane*. Moscow: Russkaya Panorama.

Szulc, Tomasz & Kiński, Andrzej (2016). "Wierba -nowy rosyjski przenośny przeciwlotniczy zestaw rakietowy," *Wojsko i Technika* 1/2016: 24–30.

Tkachev, Viktor (2005). "Strely i shilki v boyu," *Voenno-Kosmicheskaya Oborona* 6 (25): 27–32.

Transue, J.R. *et al.* (1974). *Assessment of the Weapons and Tactics used in the October 1973 War in the Mid-East*. Arlington, VA: Institute for Defense Analyses.

Vinogradov, V.M. *et al.* (1995). *Grif sekretno snyat: Kniga ob uchastii sovetskikh voennosluzhashchikh v arabo-izrailskom konflikte*. Moscow: Lespromekonomika.

Volodko, A.M. & Gorshkov, V.A. (1993). *Vertolet v Afghanistane*. Moscow: Voenizdat.

Zeigler, Sean, *et al.* (2019). *Acquisition and Use of MANPADS against Commercial Aviation: Risks, Proliferation, Mitigation, and the Cost of an Attack*. Santa Monica, CA: Rand Corporation.

INDEX